London Mathematical Society Lecture Note Series. 146

Cohen-Macaulay Modules over Cohen-Macaulay Rings

Yuji Yoshino
Yoshida College, Kyoto University

CAMBRIDGE UNIVERSITY PRESS
Cambridge
New York Port Chester Melbourne Sydney

Published by the Press Syndicate of the University of Cambridge
The Pitt Building, Trumpington Street, Cambridge CB2 1RP
40 West 20th Street, New York, NY 10011, USA
10, Stamford Road, Oakleigh, Melbourne 3166, Australia

© Cambridge University Press 1990

First published 1990

Printed in Great Britain at the University Press, Cambridge

Library of Congress cataloguing in publication data available

British Library cataloguing in publication data available

ISBN 0 521 35694 6

LONDON MATHEMATICAL SOCIETY LECTURE NOTE SERIES

Managing Editor: Professor J.W.S. Cassels, Department of Pure Mathematics and Mathematical Statistics, University of Cambridge, 16 Mill Lane, Cambridge CB2 1SB, England

The books in the series listed below are available from booksellers, or, in case of difficulty, from Cambridge University Press.

34   Representation theory of Lie groups, M.F. ATIYAH *et al*
36   Homological group theory, C.T.C. WALL (ed)
39   Affine sets and affine groups, D.G. NORTHCOTT
40   Introduction to H_p spaces, P.J. KOOSIS
43   Graphs, codes and designs, P.J. CAMERON & J.H. VAN LINT
45   Recursion theory: its generalisations and applications, F.R. DRAKE & S.S. WAINER (eds)
46   p-adic analysis: a short course on recent work, N. KOBLITZ
49   Finite geometries and designs, P. CAMERON, J.W.P. HIRSCHFELD & D.R. HUGHES (eds)
50   Commutator calculus and groups of homotopy classes, H.J. BAUES
51   Synthetic differential geometry, A. KOCK
57   Techniques of geometric topology, R.A. FENN
59   Applicable differential geometry, M. CRAMPIN & F.A.E. PIRANI
62   Economics for mathematicians, J.W.S. CASSELS
66   Several complex variables and complex manifolds II, M.J. FIELD
69   Representation theory, I.M. GELFAND *et al*
74   Symmetric designs: an algebraic approach, E.S. LANDER
76   Spectral theory of linear differential operators and comparison algebras, H.O. CORDES
77   Isolated singular points on complete intersections, E.J.N. LOOIJENGA
78   A primer on Riemann surfaces, A.F. BEARDON
79   Probability, statistics and analysis, J.F.C. KINGMAN & G.E.H. REUTER (eds)
80   Introduction to the representation theory of compact and locally compact groups, A. ROBERT
81   Skew fields, P.K. DRAXL
82   Surveys in combinatorics, E.K. LLOYD (ed)
83   Homogeneous structures on Riemannian manifolds, F. TRICERRI & L. VANHECKE
86   Topological topics, I.M. JAMES (ed)
87   Surveys in set theory, A.R.D. MATHIAS (ed)
88   FPF ring theory, C. FAITH & S. PAGE
89   An F-space sampler, N.J. KALTON, N.T. PECK & J.W. ROBERTS
90   Polytopes and symmetry, S.A. ROBERTSON
91   Classgroups of group rings, M.J. TAYLOR
92   Representation of rings over skew fields, A.H. SCHOFIELD
93   Aspects of topology, I.M. JAMES & E.H. KRONHEIMER (eds)
94   Representations of general linear groups, G.D. JAMES
95   Low-dimensional topology 1982, R.A. FENN (ed)
96   Diophantine equations over function fields, R.C. MASON
97   Varieties of constructive mathematics, D.S. BRIDGES & F. RICHMAN
98   Localization in Noetherian rings, A.V. JATEGAONKAR
99   Methods of differential geometry in algebraic topology, M. KAROUBI & C. LERUSTE
100 Stopping time techniques for analysts and probabilists, L. EGGHE

101	Groups and geometry, ROGER C. LYNDON
103	Surveys in combinatorics 1985, I. ANDERSON (ed)
104	Elliptic structures on 3-manifolds, C.B. THOMAS
105	A local spectral theory for closed operators, I. ERDELYI & WANG SHENGWANG
106	Syzygies, E.G. EVANS & P. GRIFFITH
107	Compactification of Siegel moduli schemes, C-L. CHAI
108	Some topics in graph theory, H.P. YAP
109	Diophantine Analysis, J. LOXTON & A. VAN DER POORTEN (eds)
110	An introduction to surreal numbers, H. GONSHOR
111	Analytical and geometric aspects of hyperbolic space, D.B.A.EPSTEIN (ed)
112	Low-dimensional topology and Kleinian groups, D.B.A. EPSTEIN (ed)
113	Lectures on the asymptotic theory of ideals, D. REES
114	Lectures on Bochner-Riesz means, K.M. DAVIS & Y-C. CHANG
115	An introduction to independence for analysts, H.G. DALES & W.H. WOODIN
116	Representations of algebras, P.J. WEBB (ed)
117	Homotopy theory, E. REES & J.D.S. JONES (eds)
118	Skew linear groups, M. SHIRVANI & B. WEHRFRITZ
119	Triangulated categories in the representation theory of finite-dimensional algebras, D. HAPPEL
121	Proceedings of *Groups - St Andrews 1985*, E. ROBERTSON & C. CAMPBELL (eds)
122	Non-classical continuum mechanics, R.J. KNOPS & A.A. LACEY (eds)
124	Lie groupoids and Lie algebroids in differential geometry, K. MACKENZIE
125	Commutator theory for congruence modular varieties, R. FREESE & R. MCKENZIE
126	Van der Corput's method for exponential sums, S.W. GRAHAM & G. KOLESNIK
127	New directions in dynamical systems, T.J. BEDFORD & J.W. SWIFT (eds)
128	Descriptive set theory and the structure of sets of uniqueness, A.S. KECHRIS & A. LOUVEAU
129	The subgroup structure of the finite classical groups, P.B. KLEIDMAN & M.W.LIEBECK
130	Model theory and modules, M. PREST
131	Algebraic, extremal & metric combinatorics, M-M. DEZA, P. FRANKL & I.G. ROSENBERG (eds)
132	Whitehead groups of finite groups, ROBERT OLIVER
133	Linear algebraic monoids, MOHAN S. PUTCHA
134	Number theory and dynamical systems, M. DODSON & J. VICKERS (eds)
135	Operator algebras and applications, 1, D. EVANS & M. TAKESAKI (eds)
136	Operator algebras and applications, 2, D. EVANS & M. TAKESAKI (eds)
137	Analysis at Urbana, I, E. BERKSON, T. PECK, & J. UHL (eds)
138	Analysis at Urbana, II, E. BERKSON, T. PECK, & J. UHL (eds)
139	Advances in homotopy theory, S. SALAMON, B. STEER & W. SUTHERLAND (eds)
140	Geometric aspects of Banach spaces, E.M. PEINADOR and A. RODES (eds)
141	Surveys in combinatorics 1989, J. SIEMONS (ed)
142	The geometry of jet bundles, D.J. SAUNDERS
143	The ergodic theory of discrete groups, PETER J. NICHOLLS
144	Uniform spaces, I.M. JAMES
145	Homological questions in local algebra, JAN R. STROOKER
146	Cohen-Macaulay modules over Cohen-Macaulay rings, Y. YOSHINO
147	Continuous and discrete modules, S.H. MOHAMED & B.J. MÜLLER
148	Helices and vector bundles, A.N. RUDAKOV et al
152	Oligomorphic permutation groups, P. CAMERON
154	Number theory and cryptography, J. LOXTON (ed)

Dedicated to Professor Hideyuki Matsumura on his sixtieth birthday.

Contents

Chapter 1.	Preliminaries	1
Chapter 2.	AR sequences and irreducible morphisms	8
Chapter 3.	Isolated singularities	16
Chapter 4.	Auslander categories	25
Chapter 5.	AR quivers	35
Chapter 6.	The Brauer-Thrall theorem	44
Chapter 7.	Matrix factorizations	54
Chapter 8.	Simple singularities	60
Chapter 9.	One-dimensional CM rings of finite representation type	69
Chapter 10.	McKay graphs	85
Chapter 11.	Two-dimensional CM rings of finite representation type	98
Chapter 12.	Knörrer's periodicity	106
Chapter 13.	Grothendieck groups	118
Chapter 14.	CM modules on quadrics	124
Chapter 15.	Graded CM modules on graded CM rings	135
Chapter 16.	CM modules on toric singularities	143
Chapter 17.	Homogeneous CM rings of finite representation type	161
	Addenda	168
	References	170

Preface

This is a widely revised version of lectures I gave at Tokyo Metropolitan University in 1987, originally written in Japanese [67].

Throughout the book our attention is directed to the point — how we can classify CM modules over a given CM ring, or classify CM rings which have essentially a finite number of CM modules? Being analogous to lattices over orders, this question seems to have arised very naturally. The first approach to CM modules in this direction was, perhaps, done by Herzog [35], and it turned out that there are, in themselves, two basic aspects of this problem. The first is an algebraic or representation-theoretic side, in which Auslander and Reiten made remarkable progress by the powerful use of AR sequences. The second is a geometric side, more precisely, the spectra of CM local rings having only a finite number of CM modules should be well-behaved singularities. Artin, Verdier, Knörrer, Buchweitz, Eisenbud, Greuel and many others are concerned with this direction.

In this book I have tried to give a systematic treatment of the subject, as self-contained as possible, but there is no intention to make this book an encyclopedia of CM modules. Therefore, even in the case that more general treatments for definitions or proofs exist, I have prefered to give direct expositions, which, I am afraid, the experienced reader might feel unwise of me.

I conclude this preface with acknowledgements and thanks to all who supported the preparation of the book. Especially I would like to express my great gratitude to Prof. Maurice Auslander who gave me valuable comments on the first draft. I am also indebted to Eduardo Marcos, Akira Ooishi and Kazuhiko Kurano who read the manuscript with great care and attention. Finally I wish to thank Mr. David Tranah, an editor of Cambridge U.P., for his advice in correcting my English.

January, 1990   *Yuji Yoshino*

Chapter 1. Preliminaries

In this chapter we will review some basic facts without proofs and give some of the basic notation that will be used throughout the book. For further results in commutative algebra we refer the reader to the excellent textbooks of Matsumura [47], [48] and Nagata [50]. For material such as local cohomologies and canonical modules, we recommend Herzog and Kunz [37].

Throughout this chapter R is a commutative Noetherian local ring with maximal ideal $\mathfrak{m}$ and with residue field $k = R/\mathfrak{m}$. We always denote the Krull dimension of R by d. All modules considered here will be finitely generated and unitary.

A. CM modules.

Let M be an R-module. Recall that a sequence $\{x_1, x_2, \ldots, x_n\}$ of elements in $\mathfrak{m}$ is a regular sequence on M if x_{i+1} is a non zero divisor on $M/(x_1, x_2, \ldots, x_i)M$ for any i $(0 \leq i < n)$. The depth of M is the maximum length of regular sequences on M.

In this book we shall be concerned exclusively with Cohen-Macaulay modules, which are defined as follows:

(1.1) DEFINITION. An R-module M is called a **maximal Cohen-Macaulay module** or simply a **Cohen-Macaulay** (abbr. **CM**) **module** if the depth of M is equal to d. The ring R is a CM ring if R is a CM module over R.

The reader may recall several equivalent definitions of CM modules.

(1.2) PROPOSITION. (Grothendieck [33] or Herzog-Kunz [37]) *The following conditions are equivalent for an R-module M:*
(1.2.0) M is a CM module over R;
(1.2.1) $\operatorname{Ext}^i_R(k, M) = 0$ $(i < d)$;
(1.2.2) $\operatorname{H}^i_{\mathfrak{m}}(M) = 0$ $(i \neq d)$,
where $\operatorname{H}^i_{\mathfrak{m}}$ denotes the i-th local cohomology functor with support on $\{\mathfrak{m}\}$.

In practice the conditions (1.2.1) and (1.2.2) will be very useful because of their homological nature. The next two propositions, for example, are proved by using them.

(1.3) PROPOSITION. *Let $0 \to L \to M \to N \to 0$ be an exact sequence of R-modules. Then the following hold:*
(1.3.1) *If L and N are CM, then so is M.*
(1.3.2) *If M and N are CM, then so is L.*

Warning and Exercise: It is not necessarily true that if L and M are CM, then so is N. Give a counter-example to this.

(1.4) PROPOSITION. *Let R be a CM local ring and let*

$$0 \longrightarrow M \longrightarrow F_{n-1} \longrightarrow F_{n-2} \longrightarrow \cdots \longrightarrow F_1 \longrightarrow F_0$$

be an exact sequence of R-modules where each F_i is finitely generated free. If $n \geq d$, then M is a CM module.

The following facts are rather well known and will be useful later.

(1.5) PROPOSITION.
(1.5.1) *If R is a regular local ring, then any CM module over R is a free module.*
(1.5.2) *If R is a reduced local ring of dimension 1, then an R-module M is CM only when it is torsion free, that is, when the natural homomorphism $M \to \mathrm{Hom}_R(\mathrm{Hom}_R(M,R), R)$ is a monomorphism.*
(1.5.3) *If R is a normal local domain of dimension 2, then an R-module M is CM only when it is reflexive, that is, when the natural homomorphism $M \to \mathrm{Hom}_R(\mathrm{Hom}_R(M,R), R)$ is an isomorphism.*
(1.5.4) *If R is a normal local domain of dimension ≥ 3, then any CM modules over R are reflexive. It is, however, not necessarily true that reflexive modules are CM.*

B. Multiplicities.

We now summarize some of the basic results from the theory of multiplicities. For an R-module M, it is known that the length of $M/\mathfrak{m}^n M$ is a polynomial in n if n is large enough, and the polynomial is of the form

$$(e(M)/d!)n^d + \text{(terms of degree less than } d),$$

where $e(M)$ is the **multiplicity** of M. It is true that $e(M)$ is always a nonnegative integer and that $e(M) = 0$ if and only if the Krull dimension $\dim(M)$ is less than d. Multiplicities have the property of additivity in the following sense.

(1.6) PROPOSITION. (Nagata [50, Chapter 3])
(1.6.1) *If $0 \to L \to M \to N \to 0$ is an exact sequence of R-modules, then the equality $e(M) = e(L) + e(N)$ holds.*
(1.6.2) *If R is an integral domain, then we have $e(M) = e(R) \cdot \mathrm{rank}(M)$ for any R-module M.*

(1.7) PROPOSITION. (Nagata [50, Chapter 3]) *Let M be a CM module over a local ring R. Then, for a system of parameters $\{x_1, x_2, \ldots, x_d\}$ for R, we have the inequality*

$$e(M) \leq length(M/(x_1, x_2, \ldots, x_d)M) \leq n^d \cdot e(M),$$

where n is the least integer with the property $\mathfrak{m}^n \subset (x_1, x_2, \ldots, x_d)R$.

C. *Noetherian normalization.*

Later in this book we will have occasion to encounter the case where the local ring R is an algebra over another local ring T. In such a case we have the invariance of the CM property.

(1.8) PROPOSITION. (Grothendieck [33, Cor.5.7]) *Suppose R is a finite T-algebra where T is also a local ring with the same dimension d. Then an R-module M is CM over R if and only if it is CM over T.*

In fact this proposition can be proved by using (1.2). Under the same assumption as in (1.8), the natural ring homomorphism $T \to R$ is called a **Noetherian normalization** of R when T is regular. It is a classical result that Noetherian normalizations of R exist if R is a complete local domain (Matsumura [47, (28.P)]). Combining (1.8) with (1.5.1) we have:

(1.9) PROPOSITION. *Let $T \to R$ be a Noetherian normalization of R. Then an R-module M is CM over R only when it is free regarded as a T-module.*

D. *Local duality and canonical modules.*

There is a powerful theorem called local duality for CM modules over CM rings. Before stating it we recall the definition of canonical modules.

(1.10) DEFINITION. A module K_R over a CM ring R is a **canonical module** of R if the following two conditions are satisfied:

(1.10.1) K_R is a CM module, and

(1.10.2) $\mathrm{Ext}^i_R(k, K_R) \simeq \begin{cases} 0 & (i \neq d), \\ k & (i = d). \end{cases}$

It is known that this definition is equivalent to the following single equality (Herzog-Kunz [37]).

(1.10.3) $\quad \text{Hom}_R(\text{H}^d_{\mathfrak{m}}(R), E_R(k)) \simeq \widehat{K_R}$,

where $E_R(k)$ is the injective envelope of an R-module k, and $\widehat{K_R}$ denotes the completion of K_R with respect to the $\mathfrak{m}$-adic topology.

In general a canonical module need not exist, but if it does, it is unique up to isomorphism. Fortunately we know that most rings possess canonical modules.

(1.11) PROPOSITION. (Herzog-Kunz [37]) *Suppose a CM local ring R has a Noetherian normalization $T \to R$. Then the R-module $K_R = \text{Hom}_T(R, T)$ is a canonical module of R.*

We are now ready to state the theorem of local duality.

(1.12) PROPOSITION. (Grothendieck [33, Theorem 6.3]) *Suppose a CM local ring R has the canonical module K_R. Then, for any R-module M, there are natural isomorphisms*

$$\text{Ext}^i_R(M, K_R)\widehat{} \simeq \text{Hom}_R(\text{H}^{d-i}_{\mathfrak{m}}(M), E_R(k)),$$

for any i.

As an easy consequence of this we obtain the following:

(1.13) COROLLARY. *Let M be a CM module over a CM ring R having the canonical module K_R. Then $\text{Hom}_R(M, K_R)$ is also a CM module and there is an isomorphism $M \simeq \text{Hom}_R(\text{Hom}_R(M, K_R), K_R)$. Moreover each $\text{Ext}^i_R(M, K_R)$ vanishes unless $i = 0$. In particular $\text{Hom}_R(\ , K_R)$ is an exact auto-functor on the category of CM modules over R.*

For later use we make the following remark which is also an easy corollary of (1.12).

(1.14) REMARK. Let R be the same as in (1.12) and let $0 \to K_R \to L \to M \to 0$ be an exact sequence of CM R-modules where K_R is the canonical module of R. Then the sequence splits.

Actually this is the consequence of the fact that $\text{Ext}^1_R(M, K_R) = 0$. In other words, *the canonical module is an injective object in the category of CM modules.*

E. *Syzygies.*

(1.15) DEFINITION. Consider an exact sequence of R-modules;

$$0 \longrightarrow N \longrightarrow F_{n-1} \longrightarrow F_{n-2} \longrightarrow \cdots \longrightarrow F_1 \longrightarrow F_0 \longrightarrow M \longrightarrow 0,$$

where each F_i is a free R-module. Then N is called an n-th **syzygy** of M. The **reduced** n-th syzygy $\text{syz}^n(M)$ of M is the module obtained from N by avoiding all free direct summands. Thus $\text{syz}^n(M)$ has no free direct summand and $N \simeq \text{syz}^n(M) \oplus F$ for some free module F. Notice that the reduced n-th syzygy of M is uniquely determined by M and n up to isomorphism. Note that, by definition, if M is free, then $\text{syz}^n(M) = 0$ $(n \geq 0)$. Note also that if M has projective dimension p, then $\text{syz}^n(M) = 0$ for any $n \geq p$.

Proposition (1.4) can be put in the following form in terms of syzygies.

(1.16) PROPOSITION. *Let R be a CM local ring of dimension d. Then for any R-module M and for any integer n that is not less than d, $\text{syz}^n(M)$ is either a CM module or a null module.*

This provides a possible way of constructing new CM modules by taking syzygies or reduced syzygies.

F. Henselian rings.

(1.17) DEFINITION. A local ring R is a **Henselian ring** if the following condition is satisfied.

(1.17.1) Any commutative R-algebra which is module-finite over R is a direct product of local R-algebras.

Recall that an R-module is called **indecomposable** if it has no nontrivial direct summands. There is a crucial fact concerning indecomposability of modules over a Henselian ring.

(1.18) PROPOSITION. *Let R be a Henselian local ring and let M be an R-module. Then M is indecomposable if and only if the endomorphism ring $\text{End}_R(M)$ is a local algebra, that is, sums of nonunits in $\text{End}_R(M)$ are nonunits. This assures us that the category of finitely generated R-modules admits the Krull-Schmidt theorem. Namely, any R-module is uniquely a finite direct sum of indecomposable R-modules.*

A famous theorem of Hensel asserts that complete local rings are Henselian rings. More generally, analytic algebras defined below are also Henselian.

(1.19) DEFINITION. Let k be a valued field. Thus there is a mapping v from k to the set of nonnegative real numbers, which satisfies the conditions:

(1.19.1) $v(x) = 0$ if and only if $x = 0$. And $v(xy) = v(x) \cdot v(y)$, $v(x+y) \leq v(x) + v(y)$ for any $x, y \in k$.

Then consider a formal power series f over k in n variables $\{x_1, x_2, \ldots, x_n\}$ with the following condition:

(1.19.2) Write f as $\sum a_{i_1 i_2 \ldots i_n} x_1^{i_1} x_2^{i_2} \ldots x_n^{i_n}$ ($a_{i_1 i_2 \ldots i_n} \in k$). Then there are positive real numbers $r_1, r_2, \ldots, r_n$ and N such that $v(a_{i_1 i_2 \ldots i_n}) r_1^{i_1} r_2^{i_2} \ldots r_n^{i_n} \leq N$ for all $i_1, i_2, \ldots, i_n$.

Call such series a **convergent power series** with respect to the valuation v. We denote by $k\{x_1, x_2, \ldots, x_n\}$ the set of all convergent power series, and call it a **convergent power series ring**. Notice that if the valuation is trivial, i.e. $v(x) = 1$ for any $x \neq 0$, then all formal power series are convergent ones, thus $k\{x_1, x_2, \ldots, x_n\}$ is the formal power series ring.

An **analytic algebra** over k is defined to be a finite algebra over a convergent power series ring. Any complete local ring containing a field is an analytic algebra with trivial valuation. It is known that a local analytic algebra is a Henselian ring (Nagata [50, Chapter 7]). If k is a perfect field, then every local analytic algebra is a homomorphic image of a convergent power series ring over k (Scheja-Storch [57]). This, however, is not true unless k is perfect. It is also known by Scheja-Storch [57, (8.10)] that all analytic algebras are excellent rings.

(1.20) DEFINITION AND PROPOSITION. (Scheja-Storch [57]) *Let R be a local analytic algebra over a valued field k. Then a system of parameters $\{x_1, x_2, \ldots, x_d\}$ for R is called* **separable** *if the total quotient ring of R is a separable algebra over the quotient field of the convergent power series ring $k\{x_1, x_2, \ldots, x_d\}$. If k is a perfect field, then every reduced analytic algebra over k has a separable system of parameters.*

G. Split morphisms.

We end this chapter by making several remarks about split morphisms. Recall first its definition. A homomorphism $f : M \to N$ of R-modules is a **split epimorphism** if it has a right inverse, that is, if there is a morphism g from N to M such that $f \cdot g = 1_N$. Similarly f is a **split monomorphism** if f has a left inverse. Notice that if f is a split epimorphism (resp. a split monomorphism), then N (resp. M) is isomorphic to a direct summand of M (resp. N).

For later use we remark the following:

(1.21) PROPOSITION. *Let R be a Henselian local ring and let $f : M \to L$ and $g : N \to L$ be homomorphisms of R-modules which are not split epimorphisms. Suppose L is an indecomposable R-module. Then the homomorphism $(f, g) : M \oplus N \to L$ is not a split epimorphism.*

This is evident from (1.18). Actually, if (f, g) were a split epimorphism, then there would be $a : L \to M$ and $b : L \to N$ satisfying $f \cdot a + g \cdot b = 1_L$. Since $\mathrm{End}_R(L)$ is local, this would imply that either $f \cdot a$ or $g \cdot b$ is an automorphism on L, which is a contradiction.

The following is also easy to see.

(1.22) PROPOSITION. *Assume that $f : N \to M$ is a homomorphism of R-modules. Let $N = \sum_i N_i$ be a direct decomposition of N into indecomposable modules and let $M_j = M/f(\sum_{i \neq j} N_i)$. Consider R-homomorphisms $f_j : N_j \to M_j$ for any j which are induced naturally from f. If all of the f_j are split monomorphisms, then so is f.*

Chapter 2. AR sequences and irreducible morphisms

This chapter introduces some of the basic theory of AR sequences, which will play the key role in later part of this book. Auslander and Reiten introduced this notion for their theory of representations of Artinian algebras. They, and several others, developed the theory to much wider classes of categories, including the category of CM modules, see [5] for instance. In what follows, 'AR' always stands for 'Auslander and Reiten'.

In this chapter R is always a Henselian CM local ring with maximal ideal $\mathfrak{m}$ and with residue field k. We always denote the Krull dimension of R by d. It is convenient now to introduce the notation for categories of modules. The category of all finitely generated R-modules and R-homomorphisms will be denoted by $\mathfrak{M}(R)$. The full subcategory of $\mathfrak{M}(R)$ consisting of all CM modules will be denoted by $\mathfrak{C}(R)$. Notice that $M \in \mathfrak{C}(R)$ is indecomposable if and only if $\operatorname{End}_R(M)$ is a local ring, cf. (1.18).

Before giving a precise definition of AR sequences it is necessary to introduce a further notion.

(2.1) DEFINITION. For an indecomposable CM module $M \in \mathfrak{C}(R)$, we define a set of short exact sequences $S(M)$ as follows:

$$S(M) = \{s : 0 \to N_s \to E_s \to M \to 0 \mid$$
$$s \text{ is a } \textit{nonsplit} \text{ exact sequence in } \mathfrak{C}(R) \text{ with } N_s \text{ indecomposable}\}.$$

In particular, an element of $S(M)$ gives a nontrivial element of $\operatorname{Ext}^1_R(M, N_s)$.

The next is a direct consequence of the definition.

(2.2) LEMMA. *If M is an indecomposable CM module over R which is not free, then $S(M)$ is nonempty.*

PROOF: Let s be a nonsplit exact sequence $0 \to N \to E \to M \to 0$ in $\mathfrak{C}(R)$ which ends in M. Since M is nonfree, there exists at least one such exact sequence. For example, it is enough to take E as a free cover of M. Decompose N into indecomposable modules as $N = \sum_i N_i$ and let E_j be $E / \sum_{i \neq j} N_i$. Consider new exact sequences $s_j : 0 \to N_j \to E_j \to M \to 0$. Since s is nonsplit, one of the sequences s_j is also nonsplit (1.22) and thus it lies in $S(M)$. ∎

(2.3) DEFINITION. Let s and t be two elements of $S(M)$.
(2.3.1) We write $s > t$ if there is an $f \in \operatorname{Hom}_R(N_s, N_t)$ such that $\operatorname{Ext}_R^1(M, f)(s) = t$. In this case we say that s is *bigger* than t or t is *smaller* than s. This is equivalent to the existence of a commutative diagram:

$$\begin{array}{ccccccccc} 0 & \to & N_s & \to & E_s & \to & M & \to & 0 \\ & & f\downarrow & & \downarrow & & \| & & \\ 0 & \to & N_t & \to & E_t & \to & M & \to & 0. \end{array}$$

(2.3.2) We write $s \sim t$ if f is an isomorphism above in (2.3.1). We often identify s with t when $s \sim t$.

(2.4) LEMMA. *Let M be an indecomposable CM module and let s and t be in $S(M)$. If $s > t$ and $t > s$, then we have $s \sim t$. In particular $S(M)$ is a well-defined partially ordered set.*

PROOF: There are, by definition, R-homomorphisms $f : N_s \to N_t$ and $g : N_t \to N_s$ satisfying $\operatorname{Ext}_R^1(M, f)(s) = t$ and $\operatorname{Ext}_R^1(M, g)(t) = s$. If we denote the composition $g \cdot f$ by h, then we have $\operatorname{Ext}_R^1(M, h)(s) = s$. Since both N_s and N_t are indecomposable, it is enough to show that h is an isomorphism. Thus the lemma follows from the following:

(2.5) LEMMA. *Let s be an element of $S(M)$ and let h be an endomorphism of N_s. If $\operatorname{Ext}_R^1(M, h)(s) = s$, then h is an automorphism of N_s.*

PROOF: Suppose h is not an isomorphism. Then h belongs to the Jacobson radical of $\operatorname{End}_R(N_s)$ and hence some power of h is in $\mathfrak{m} \operatorname{End}_R(N_s)$. We may thus assume that h itself is in $\mathfrak{m} \operatorname{End}_R(N_s)$. Then, for any integer n, we have $h^n = \sum_i a_{in} g_{in}$ for some $a_{in} \in \mathfrak{m}^n$ and $g_{in} \in \operatorname{End}_R(N_s)$. Therefore $s = \operatorname{Ext}_R^1(M, h^n)(s) = \sum_i a_{in} \operatorname{Ext}_R^1(M, g_{in})(s)$ is in $\mathfrak{m}^n \operatorname{Ext}_R^1(M, N_s)$ for any n. Thus we conclude that $s = 0$ as an element of $\operatorname{Ext}_R^1(M, N_s)$ which is clearly a contradiction, for s is nonsplit. ∎

The partially ordered set $S(M)$ has the following property:

(2.6) LEMMA. *Let M be an indecomposable CM module and let $s : 0 \to N_s \to E_s \xrightarrow{p} M \to 0$ and $t : 0 \to N_t \to E_t \xrightarrow{q} M \to 0$ be in $S(M)$. Then there is an element $u \in S(M)$ such that $s > u$ and $t > u$.*

PROOF: Consider an exact sequence $0 \to N \to E \xrightarrow{\varphi} M \to 0$, where $E = E_s \oplus E_t$ and N is the kernel of the homomorphism $\varphi = (p, q)$. Decompose N into indecomposable modules as $N = \sum_i N_i$ and denote $E_j = E/\sum_{i \neq j} N_i$. Then we know from (1.22) that one of the sequences $u_j : 0 \to N_j \to E_j \to M \to 0$ is in $S(M)$, say $u_1 \in S(M)$. Then, from the definition, it is easy to see that $s > u_1$ and $t > u_1$. ∎

(2.7) COROLLARY. *Let M be an indecomposable CM module. If s is a minimal element in $S(M)$, then it is minimum in $S(M)$.*

We are now ready to define AR sequences.

(2.8) DEFINITION. *Let M be an indecomposable CM module over R. A short exact sequence $s : 0 \to N_s \to E_s \to M \to 0$ is an **AR sequence ending in** M if s is the minimum element in $S(M)$.* An AR sequence ending in M is, if it exists, uniquely determined by M. In particular the modules N_s and E_s are also unique up to an isomorphism. If s is the AR sequence ending in M, then we denote N_s by $\tau(M)$ and call it the **AR translation** of M.

This definition of AR sequences looks very theoretical. We shall rewrite it for practical use.

(2.9) LEMMA. *Let M be an indecomposable CM module and let $s : 0 \to N_s \to E_s \xrightarrow{p} M \to 0$ be in $S(M)$. Then the following two conditions are equivalent:*
(2.9.1)   *s is the AR sequence ending in M;*
(2.9.2)   *For any R-homomorphism $q : L \to M$ in $\mathfrak{C}(R)$ which is not a split epimorphism, there is an R-homomorphism $f : L \to E_s$ such that $q = p \cdot f$.*

PROOF: (2.9.2) $\Rightarrow$ (2.9.1)   Let $t : 0 \to N_t \to E_t \xrightarrow{q} M \to 0$ be a sequence in $S(M)$ with $t < s$. We want to show that $s < t$. Since q is not a split epimorphism, we know from (2.9.2) that there is an $f : E_t \to E_s$ satisfying $q = p \cdot f$. Denote by $g : N_t \to N_s$ the restriction of f on N_t. Then it follows that $\mathrm{Ext}^1_R(M, g)(t) = s$, that is, $s < t$.

(2.9.1) $\Rightarrow$ (2.9.2)   Let $q : L \to M$ be a homomorphism which is not a split epimorphism. We construct a new exact sequence

$$u : 0 \longrightarrow Q \longrightarrow E_s \oplus L \xrightarrow{\varphi} M \longrightarrow 0,$$

where φ denotes the homomorphism (p, q) and Q is the kernel of φ. Since both p and q are nonsplit, the sequence u is also nonsplit by (1.21). Denoting by h the restriction of the natural monomorphism $E_s \to E_s \oplus L$ on N_s, we see that h is a homomorphism from N_s into Q with the property $\mathrm{Ext}^1_R(M, h)(s) = u$. Decompose Q into indecomposable modules to write $Q = \sum_i Q_i$, hence $\mathrm{Ext}^1_R(M, Q) = \sum_i \mathrm{Ext}^1_R(M, Q_i)$. Write $u = \sum_i u_i$ along this decomposition. Since $u \neq 0$ in $\mathrm{Ext}^1_R(M, Q)$, one of the u_i is nonsplit. Denote it by t. It can be seen from the definition that t is an element of $S(M)$ and that $s > t$. Then from the assumption (2.9.1) we see that $s \sim t$. In particular there is a homomorphism $g : Q \to N_s$ which makes the following diagram commutative:

$$\begin{array}{ccccccccc} 0 & \longrightarrow & Q & \longrightarrow & E_s \oplus L & \longrightarrow & M & \longrightarrow & 0 \\ & & {\scriptstyle g}\downarrow & & {\scriptstyle f'}\downarrow & & \parallel & & \\ 0 & \longrightarrow & N_s & \longrightarrow & E_s & \longrightarrow & M & \longrightarrow & 0 \end{array}$$

Denoting by f the composition of the homomorphism f' appearing in the middle of the diagram with the natural monomorphism $L \to E_s \oplus L$, we obtain $q = p \cdot f$. This completes the proof. ∎

Auslander and Reiten, who initiated the theory of AR sequences, called an AR sequence an 'almost split' sequence, which is apparently named after the property (2.9.2); see Auslander-Reiten [9].

We next make a very important definition.

(2.10) DEFINITION. Let M and N be CM modules over R and let $f : M \to N$ be an R-homomorphism. Call f an **irreducible morphism** if the following two conditions are satisfied:

(2.10.1) f is neither a split epimorphism nor a split monomorphism.

(2.10.2) If a commutative diagram

$$\begin{array}{ccc} M & \xrightarrow{f} & N \\ {\scriptstyle g}\searrow & & \nearrow{\scriptstyle h} \\ & X & \end{array}$$

in $\mathfrak{C}(R)$ is given, then either g is a split monomorphism or h is a split epimorphism.

(2.11) LEMMA. *Let M be an indecomposable CM module over R and let $s : 0 \to N \to E \xrightarrow{p} M \to 0$ be an element of $S(M)$. Suppose that s is the AR sequence ending in M. Then p is an irreducible morphism.*

PROOF: It suffices to check that the homomorphism p satisfies the condition (2.10.2). So consider a commutative diagram

$$\begin{array}{ccc} E & \xrightarrow{p} & M \\ {\scriptstyle g}\searrow & & \nearrow{\scriptstyle h} \\ & X & \end{array}$$

in $\mathfrak{C}(R)$ and assume that h is not a split epimorphism. It is enough to see that g is a split monomorphism. Let $G = E \oplus X$ and let φ be a homomorphism $(p, h) : G \to M$.

Then we have the commutative diagram:

$$\begin{array}{ccccccccc} 0 & \longrightarrow & N & \longrightarrow & E & \stackrel{p}{\longrightarrow} & M & \longrightarrow & 0 \\ & & \downarrow & & j\downarrow & & \| & & \\ 0 & \longrightarrow & Q & \longrightarrow & G & \stackrel{\varphi}{\longrightarrow} & M & \longrightarrow & 0, \end{array}$$

where Q is the kernel of φ and $j = \binom{0}{g}$. Since both p and h are nonsplit, φ is also nonsplit by (1.21). Decompose Q into indecomposables to write $Q = \sum_i Q_i$ and denote $G_i = G/\sum_{j \neq i} Q_j$. By (1.22), there is an i such that the sequence $s_i : 0 \to Q_i \to G_i \to M \to 0$ is nonsplit, hence $s_i \in S(M)$. Notice that the following commutative diagram exists:

$$\begin{array}{ccccccccc} 0 & \longrightarrow & Q & \longrightarrow & G & \stackrel{\varphi}{\longrightarrow} & M & \longrightarrow & 0 \\ & & \downarrow & & k\downarrow & & \| & & \\ 0 & \longrightarrow & Q_i & \longrightarrow & G_i & \longrightarrow & M & \longrightarrow & 0, \end{array}$$

where k is the natural projection. Combining two commutative diagrams above, we obtain $s_i < s$. Since s is the minimal element in $S(M)$, we see that $s \sim s_i$. Then we know from (2.5) that the composition $k \cdot j$ is an isomorphism. Writing $k = (a, b)$ along the decomposition $G = E \oplus X$, we conclude that $b \cdot g : E \to G_i$ is an isomorphism. In particular we have $\{(b \cdot g)^{-1}b\} \cdot g = 1_E$, and this shows that g is a split monomorphism. ∎

We immediately get:

(2.12) COROLLARY. *Let M and L be indecomposable CM modules over R and assume that there exists an AR sequence $s : 0 \to N \to E \stackrel{p}{\to} M \to 0$ ending in M. Then the following two conditions are equivalent:*
(2.12.1) *There is an irreducible morphism from L to M;*
(2.12.2) *L is isomorphic to a direct summand of E.*

PROOF: (2.12.1) ⇒ (2.12.2)   Let $f : L \to M$ be an irreducible morphism. Since f is not a split epimorphism and since s is an AR sequence, we see from (2.9) that there is a homomorphism $g : L \to E$ such that $f = p \cdot g$. It follows that g is a split monomorphism, for f is irreducible and p is not a split epimorphism.

(2.12.2) ⇒ (2.12.1)   Assume the decomposition of E is given by $E = L \oplus Q$. Denote $p = (f, g)$ along this decomposition. We show that f is actually an irreducible morphism from L to M. For this, let

$$\begin{array}{ccc} L & \stackrel{f}{\longrightarrow} & M \\ {\scriptstyle h}\searrow & & \nearrow{\scriptstyle k} \\ & X & \end{array}$$

be a commutative diagram in $\mathfrak{C}(R)$ and suppose that k is not a split epimorphism. We want to verify that h is a split monomorphism. Consider the commutative diagram

$$\begin{array}{ccc} E = L \oplus Q & \xrightarrow{p} & M \\ {}_{\theta}\searrow & \nearrow_{(k,g)} & \\ & X \oplus Q, & \end{array}$$

where $\theta = \begin{pmatrix} h & 0 \\ 0 & 1 \end{pmatrix}$. Since p is irreducible (2.11) and since (k,g) is not a split epimorphism (1.21), we see that θ is a split monomorphism, and hence h itself is a split monomorphism. ∎

As a corollary of this proof we obtain the following:

(2.13) COROLLARY. *Let M and L be indecomposable CM modules over R and assume that there exists an AR sequence $s : 0 \to N \to E \xrightarrow{p} M \to 0$ ending in M. Then any irreducible morphism g from L to M is obtained in the following way: There is a split monomorphism h from L to E such that $g = p \cdot h$.*

Recall that there is a duality in our category $\mathfrak{C}(R)$ as was seen in (1.12) and (1.13). We will, thus, have definitions and lemmas that are dual to the above. We will list them below and leave proofs to the reader as exercises.

(2.1)′ DEFINITION. For an indecomposable CM module $N \in \mathfrak{C}(R)$, we define a set of short exact sequences $S'(N)$ as follows:

$$S'(N) = \{s : 0 \to N \to G_s \to M_s \to 0 \mid$$
$$s \text{ is a nonsplit exact sequence in } \mathfrak{C}(R) \text{ with } M_s \text{ indecomposable}\}.$$

In particular, an element of $S'(N)$ gives a nontrivial element of $\mathrm{Ext}^1_R(M_s, N)$.

If the CM ring R has the canonical module K_R, and if we denote by $(\ )'$ the canonical dual (e.g. $M' = \mathrm{Hom}_R(M, K_R)$, $s' = \mathrm{Hom}_R(s, K_R)$ etc.), then a nonsplit short exact sequence $s : 0 \to N \to G \to M \to 0$ belongs to $S'(N)$ if and only if it is in $S(M)$, which is also equivalent to $s' \in S(N')$; see Corollary (1.13).

(2.2)′ LEMMA. *If N is an indecomposable CM module over R which is not isomorphic to the canonical module K_R, then $S'(N)$ is nonempty.*

(2.3)′ DEFINITION. Let s and t be two elements of $S'(N)$.
(2.3.1)′ We write $s >' t$ if there is an $f \in \mathrm{Hom}_R(M_t, M_s)$ such that $\mathrm{Ext}^1_R(f, N)(s) = t$.

This is equivalent to the existence of a commutative diagram:

$$\begin{array}{ccccccccc} 0 & \longrightarrow & N & \longrightarrow & G_s & \longrightarrow & M_s & \longrightarrow & 0 \\ & & \| & & \uparrow & & f\uparrow & & \\ 0 & \longrightarrow & N & \longrightarrow & G_t & \longrightarrow & M_t & \longrightarrow & 0. \end{array}$$

(2.3.2)' We write $s \sim' t$ if f, in (2.3.1)' above, is an isomorphism. We often identify s with t when $s \sim' t$.

If R has a canonical module and if $(\ )'$ denotes the canonical dual, then note that $s >' t$ in $S'(N)$ if and only if $s' > t'$ in $S(N')$.

(2.4)' LEMMA. *Let N be an indecomposable CM module and let s and t be in $S'(N)$. If $s >' t$ and $t >' s$, then we have $s \sim' t$. In particular $S'(N)$ is a well-defined partially ordered set.*

(2.5)' LEMMA. *Let s be in $S'(N)$ and let h be an endomorphism of M_s. If $\mathrm{Ext}^1_R(h, N)(s) = s$, then h is an automorphism of M_s.*

(2.6)' LEMMA. *Let N be a CM module and let s and t be in $S'(N)$. Then there is an element $u \in S'(N)$ such that $s >' u$ and $t >' u$.*

(2.7)' COROLLARY. *Let N be an indecomposable CM module. If s is a minimal element in $S'(N)$, then it is minimum in $S'(N)$.*

(2.8)' DEFINITION. Let N be an indecomposable CM module over R. A short exact sequence $s: 0 \to N \to G_s \to M_s \to 0$ is an **AR sequence starting from** N if s is the minimum element in $S'(N)$. An AR sequence starting from N is, if it exists, uniquely determined by N. In particular the modules M_s and G_s are also unique up to an isomorphism. If s is the AR sequence starting from N, then we often denote M_s by $\tau^{-1}(N)$.

(2.9)' LEMMA. *Let N be an indecomposable CM module and let $s: 0 \to N \xrightarrow{q} G_s \to M_s \to 0$ be an element in $S'(N)$. Then the following two conditions are equivalent:*
(2.9.1)' *s is the AR sequence starting from N;*
(2.9.2)' *For any R-homomorphism $r: N \to L$ in $\mathfrak{C}(R)$ which is not a split monomorphism, there is an R-homomorphism $f: G_s \to L$ such that $r = f \cdot q$.*

(2.11)' LEMMA. *Let N be an indecomposable CM module over R and let $s: 0 \to N \xrightarrow{q} G \to M \to 0$ be an element of $S'(N)$. Suppose that s is the AR sequence starting from N. Then q is an irreducible morphism.*

(2.12)' COROLLARY. *Let N and L be indecomposable CM modules over R and assume that there exists an AR sequence $s: 0 \to N \xrightarrow{q} G \to M \to 0$ starting from N. Then the following two conditions are equivalent:*

(2.12.1)' There is an irreducible morphism from N into L;
(2.12.2)' L is isomorphic to a direct summand of G.

(2.13)' COROLLARY. *Let N, L and s be the same as in (2.12)'. Then any irreducible morphism g from N to L is obtained in the following way: There is a split epimorphism h from G to L such that $g = h \cdot q$.*

We shall prove that two definitions of AR sequences that are dual to each other will be the same in the following sense.

(2.14) LEMMA. *Let $s : 0 \to N \to E \to M \to 0$ be a nonsplit short exact sequence in $\mathfrak{C}(R)$. Assume that both N and M are indecomposable. Then the following are equivalent:*
(2.14.1) *s is the AR sequence ending in M;*
(2.14.2) *s is the AR sequence starting from N.*

PROOF: We only prove that (2.14.1) implies (2.14.2). The reverse implication is similar and is left to the reader.

Let t be an element of $S'(N)$ with $t <' s$. We want to show that $s \sim' t$. By definition, there is a commutative diagram

$$\begin{array}{ccccccccc} 0 & \longrightarrow & N & \stackrel{q}{\longrightarrow} & E & \stackrel{p}{\longrightarrow} & M & \longrightarrow & 0 \\ & & \| & & f\uparrow & & g\uparrow & & \\ 0 & \longrightarrow & N & \stackrel{b}{\longrightarrow} & G_t & \stackrel{a}{\longrightarrow} & M_t & \longrightarrow & 0. \end{array}$$

Suppose g is not an isomorphism. Then it is not even a split epimorphism. Thus from (2.9) there is $\theta : M_t \to E$ such that $p \cdot \theta = g$. Then $f - \theta \cdot a$ induces a homomorphism φ from G_t to N satisfying $q \cdot \varphi = f - \theta \cdot a$, because $p \cdot (f - \theta \cdot a) = 0$. Thus we have $q \cdot \varphi \cdot b = f \cdot b - \theta \cdot a \cdot b = q$, therefore $\varphi \cdot b$ is the identity mapping on N. This contradicts the fact that t is nonsplit. Thus g must be an isomorphism, and hence $s \sim' t$. ∎

We end this chapter by defining what will be one of main topics of the next chapter.

(2.15) DEFINITION. We say that the category $\mathfrak{C}(R)$ **admits AR sequences** if, for any indecomposable CM module M over R which is not free, there exists an AR sequence ending in M.

If the CM ring has the canonical module, then the canonical dual gives an exact antiequivalence of the category $\mathfrak{C}(R)$ onto itself. Therefore that $\mathfrak{C}(R)$ admits AR sequences is equivalent to the existence of an AR sequence starting from N for any indecomposable CM module N which is not isomorphic to the canonical module.

Chapter 3. Isolated singularities

In this chapter we intend studying in detail the condition that the category $\mathfrak{C}(R)$ admits AR sequences. It will turn out that this is equivalent to the ring R having only an isolated singularity. This result was first proved by Auslander [7].

In this chapter R is a Henselian CM local ring with maximal ideal $\mathfrak{m}$ and with residue field $k = R/\mathfrak{m}$. We always assume that R has the canonical module K_R. As before d denotes the Krull dimension of R and $\mathfrak{C}(R)$ is the category of all CM modules over R.

We begin with the definition of isolated singularities.

(3.1) DEFINITION. The ring R is an **isolated singularity** (or R has only an isolated singularity) if the localizations $R_\mathfrak{p}$ are regular local rings for prime ideals $\mathfrak{p}$ of R which are distinct from $\mathfrak{m}$.

For example, if $d = 1$, then R is an isolated singularity if and only if it is reduced (i.e. having no nontrivial nilpotent elements). If $d = 2$, then it is equivalent to requiring R to be a normal integral domain. The reader should prove this as an exercise.

The goal of this chapter is to prove the following theorem.

(3.2) THEOREM. (Auslander [7]) *The following two conditions concerning R are equivalent:*
(3.2.1) *R is an isolated singularity;*
(3.2.2) *The category $\mathfrak{C}(R)$ admits AR sequences.*

In order to prove this, several lemmas will be necessary.

(3.3) LEMMA. *The following are equivalent:*
(3.3.1) *R is an isolated singularity;*
(3.3.2) *For any two CM modules M and N over R, $\mathrm{Ext}^1_R(M, N)$ is a module of finite length.*

(3.3.3) *Any CM module over R is locally free on the punctured spectrum of R.* (We say that M is **locally free on the punctured spectrum of** R if $M_\mathfrak{p}$ is $R_\mathfrak{p}$-free for any prime $\mathfrak{p}$ distinct from $\mathfrak{m}$.)

PROOF: (3.3.1) $\Rightarrow$ (3.3.3): Let M be a CM module over R and let $\mathfrak{p}$ be a prime ideal of R with $\mathfrak{p} \neq \mathfrak{m}$. Then $M_\mathfrak{p}$ is a CM module over $R_\mathfrak{p}$. Hence it follows from (1.5.1) that $M_\mathfrak{p}$ is $R_\mathfrak{p}$-free, for $R_\mathfrak{p}$ is regular.

(3.3.3) $\Rightarrow$ (3.3.2): Let M and N be CM modules. For a prime ideal $\mathfrak{p} \neq \mathfrak{m}$, we have, since $M_\mathfrak{p}$ is free over $R_\mathfrak{p}$, $\operatorname{Ext}^1_R(M,N)_\mathfrak{p} = \operatorname{Ext}^1_{R_\mathfrak{p}}(M_\mathfrak{p}, N_\mathfrak{p}) = 0$. Because this holds for any prime $\mathfrak{p}$ other than $\mathfrak{m}$, $\operatorname{Ext}^1_R(M,N)$ is a module of finite length.

(3.3.2) $\Rightarrow$ (3.3.1): Let $\mathfrak{p}$ be a prime ideal of R which is different from $\mathfrak{m}$. Consider a free resolution of $R/\mathfrak{p}$ to obtain an exact sequence

$$0 \longrightarrow M \longrightarrow F_{d-1} \longrightarrow F_{d-2} \longrightarrow \cdots \longrightarrow F_1 \longrightarrow R \longrightarrow R/\mathfrak{p} \longrightarrow 0,$$

where each F_i is a free module over R. Recall that M is a CM module over R; (1.4). It will be sufficient to show that $M_\mathfrak{p}$ is a free module over $R_\mathfrak{p}$. Actually if this is true, then the residue field $(R/\mathfrak{p})_\mathfrak{p}$ of $R_\mathfrak{p}$ has finite projective dimension and hence $R_\mathfrak{p}$ must be regular as required.

We now prove that $M_\mathfrak{p}$ is free. For this, consider the first syzygy of M and get an exact sequence

(*) $$0 \longrightarrow N \longrightarrow F_d \longrightarrow M \longrightarrow 0,$$

where F_d is free over R. Recall again that N is a CM module over R. Therefore it turns out from the assumption that $\operatorname{Ext}^1_{R_\mathfrak{p}}(M_\mathfrak{p}, N_\mathfrak{p}) = \operatorname{Ext}^1_R(M,N)_\mathfrak{p} = 0$, which precisely means that the localized sequence $0 \to N_\mathfrak{p} \to (F_d)_\mathfrak{p} \to M_\mathfrak{p} \to 0$ splits and so $M_\mathfrak{p}$ is free over $R_\mathfrak{p}$. ∎

By this lemma Theorem (3.2) follows from the following more general one.

(3.4) THEOREM. *For an indecomposable non-free CM module over R, the following two conditions are equivalent:*
(3.4.1) *M is locally free on the punctured spectrum of R;*
(3.4.2) *There is an AR sequence ending in M.*

PROOF OF (3.4.2) $\Rightarrow$ (3.4.1): Suppose that M is not locally free on the punctured spectrum of R, so that $M_\mathfrak{p}$ is not $R_\mathfrak{p}$-free for some $\mathfrak{p} \neq \mathfrak{m}$. Consider the first syzygy of M ; $0 \to L \to F \to M \to 0$, where F is R-free and L is CM over R. Since this is not split even when we take the localization at $\mathfrak{p}$, we see that $\operatorname{Ext}^1_R(M,L)_\mathfrak{p} \neq 0$. In particular, we have $\operatorname{Ext}^1_R(M,N)_\mathfrak{p} \neq 0$ for some indecomposable summand N of L.

Therefore we may find an element $s \in \mathrm{Ext}_R^1(M,N)$ and $r \in \mathfrak{m} - \mathfrak{p}$ such that $r^n s \neq 0$ for any $n \geq 1$. Remark that each $r^n s$ gives an element of $S(M)$. From the assumption there exists a sequence $t : 0 \to N_t \to E_t \to M \to 0$ with the property $t < r^n s$ for all n. This means that, for each n, there is a homomorphism $f_n : N \to N_t$ such that $\mathrm{Ext}_R^1(M, f_n)(r^n s) = t$. Since $\mathrm{Ext}_R^1(M, f_n)$ is an R-homomorphism of $\mathrm{Ext}_R^1(M,N)$ to $\mathrm{Ext}_R^1(M, N_t)$, it follows that $t \in r^n \mathrm{Ext}_R^1(M, N_t)$ for all n. This implies that $t = 0$, because $r \in \mathfrak{m}$ and $\cap_{n=1}^\infty \mathfrak{m}^n \mathrm{Ext}_R^1(M,N) = 0$. This is absurd, since t is nonsplit. ∎

The other part of Theorem (3.4) is much harder to prove and we need several more definitions and results.

(3.5) DEFINITION. Let M be a finitely generated module over R. Consider a finite presentation of M by free modules; $F_1 \xrightarrow{f} F_0 \to M \to 0$. Then put $\mathrm{tr}(M) = \mathrm{Coker}(\mathrm{Hom}_R(f, R))$ and call it the **Auslander transpose** of M. Note that $\mathrm{tr}(M)$ depends on the presentation of M. Actually another presentation $F_1' \xrightarrow{f'} F_0' \to M \to 0$ may give a distinct module. However, it can be easily seen that there are free modules F and G such that $\mathrm{Coker}(\mathrm{Hom}_R(f, R)) \oplus F \simeq \mathrm{Coker}(\mathrm{Hom}_R(f', R)) \oplus G$. (Prove this.) Thus $\mathrm{tr}(M)$ is unique up to free summand. Later on in this book, we shall deal only with properties that are independent of free summands of $\mathrm{tr}(M)$ and so the above 'rough' definition will be sufficient. The reader should note, for instance, that $\mathrm{Tor}_i^R(\mathrm{tr}(M), \ )$ and $\mathrm{Ext}_R^i(\mathrm{tr}(M), \ )$ are uniquely determined if $i \geq 1$.

Concerning the Auslander transpose, the following result is basic. We do not prove it, but refer the reader to Evans-Griffith [29] or Auslander-Bridger [8].

(3.6) PROPOSITION. *Suppose a finitely generated R-module M is locally free on the punctured spectrum of R. Then M is CM if and only if $\mathrm{Ext}_R^i(\mathrm{tr}(M), R) = 0$ $(1 \leq i \leq d)$.*

(3.7) DEFINITION. Let M and N be finitely generated R-modules. Denote by $\mathfrak{P}(M, N)$ the set of R-homomorphisms of M to N which pass through free modules. That is, an R-homomorphism $f : M \to N$ lies in $\mathfrak{P}(M, N)$ if and only if it is factored as $M \to F \to N$ with F free. Note that $\mathfrak{P}(M, N)$ is an R-submodule of $\mathrm{Hom}_R(M, N)$. We also denote

$$\underline{\mathrm{Hom}}_R(M, N) = \mathrm{Hom}_R(M, N)/\mathfrak{P}(M, N)$$

and write $\underline{\mathrm{End}}_R(M)$ instead of $\underline{\mathrm{Hom}}_R(M, M)$. We remark that $\underline{\mathrm{End}}_R(M)$ is a ring, for $\mathfrak{P}(M, M)$ is a two-sided ideal of $\mathrm{End}_R(M)$. Consequently, if M is indecomposable, then $\underline{\mathrm{End}}_R(M)$ is a local ring, by (1.18). Notice also that $\underline{\mathrm{Hom}}_R(M, N)$ is a right $\underline{\mathrm{End}}_R(M)$-module and a left $\underline{\mathrm{End}}_R(N)$-module by the natural right (resp. left) action of $\mathrm{End}_R(M)$ (resp. $\mathrm{End}_R(N)$) on $\mathrm{Hom}_R(M, N)$.

It is possible to characterize $\underline{\mathrm{Hom}}_R(M, N)$ in the following manner.

(3.8) LEMMA. *For finitely generated R-modules M and N, there is an exact sequence*

$$\text{Hom}_R(M, R) \otimes_R N \xrightarrow{q} \text{Hom}_R(M, N) \longrightarrow \underline{\text{Hom}}_R(M, N) \longrightarrow 0,$$

where q is defined by

$$q(f \otimes x)(y) = f(y)x \quad (x \in N, y \in M, f \in \text{Hom}_R(M, R)).$$

PROOF: It suffices to show that the image of q is exactly $\mathfrak{P}(M, N)$. Defining $i_x : R \to N$ ($x \in N$) by $i_x(r) = rx$, we see that $q(f \otimes x) = i_x \cdot f$. Since this actually goes through the free module R, we obtain $q(f \otimes x) \in \mathfrak{P}(M, N)$. Conversely, let $h \in \mathfrak{P}(M, N)$. By definition, h is a composition of two homomorphisms $a : R^{(n)} \to N$ and $b : M \to R^{(n)}$. Fixing a free basis $\{e_i | i = 1, 2, \ldots, n\}$ of $R^{(n)}$, we may write b as $\sum b_i(\ )e_i$ ($b_i \in \text{Hom}_R(M, R)$). Then it is easy to see that $h = \sum_i q(b_i \otimes a(e_i))$, therefore h lies inside the image of q. ∎

(3.9) LEMMA. *Let M and N be finitely generated R-modules. Then we have a natural isomorphism of* $\underline{\text{End}}_R(M) \times \underline{\text{End}}_R(N)$-*modules:*

$$\underline{\text{Hom}}_R(M, N) \simeq \text{Tor}_1^R(\text{tr}(M), N).$$

PROOF: Before proceeding to the proof, let us comment on the action of $\underline{\text{End}}_R(M) \times \underline{\text{End}}_R(N)$ on $\text{Tor}_1^R(\text{tr}(M), N)$. Let f be an element of $\text{End}_R(N)$. Then $\text{Tor}_1^R(\text{tr}(M), f)$ induces an endomorphism on $\text{Tor}_1^R(\text{tr}(M), N)$. Since $\text{Tor}_1^R(\text{tr}(M), f) = 0$ for any $f \in \mathfrak{P}(N, N)$, this gives a left action of $\underline{\text{End}}_R(N)$. On the other hand, if g is an element of $\text{End}_R(M)$, then g will induce a homomorphism of the presentation of M, that is, there is a commutative diagram

$$\begin{array}{ccccccc} F_1 & \longrightarrow & F_0 & \longrightarrow & M & \longrightarrow & 0 \\ \downarrow & & \downarrow & & {\scriptstyle g}\downarrow & & \\ F_1 & \longrightarrow & F_0 & \longrightarrow & M & \longrightarrow & 0, \end{array}$$

where each F_i is a free module. Therefore g induces a mapping of $\text{tr}(M)$ into itself. We denote this by $\text{tr}(g)$. We should remark that this notation has the same ambiguity as does $\text{tr}(M)$. Note, though, that $\text{Tor}_1^R(\text{tr}(g), N)$ is determined definitely by g and N, because $\text{Tor}_1^R(F, \ )$ and $\text{Tor}_1^R(h, \ )$ will vanish when F is free and h is a homomorphism passing through F. Therefore $\text{Tor}_1^R(\text{tr}(g), N)$ gives an action of $\underline{\text{End}}_R(M)$ on $\text{Tor}_1^R(\text{tr}(M), N)$.

Now we prove the lemma. Consider a presentation of M by free modules; $F_1 \xrightarrow{f} F_0 \to M \to 0$. Then, by definition, we have the following exact sequence:

$$0 \longrightarrow \operatorname{Hom}_R(M, R) \xrightarrow{j} \operatorname{Hom}_R(F_0, R) \xrightarrow{f^*} \operatorname{Hom}_R(F_1, R) \longrightarrow \operatorname{tr}(M) \longrightarrow 0.$$

Therefore, we get an isomorphism

$$\operatorname{Tor}_1^R(\operatorname{tr}(M), N) \simeq \operatorname{Ker}(f^* \otimes N)/(j \otimes 1)(\operatorname{Hom}_R(M, R) \otimes N).$$

Here, remarking that there is an isomorphism $h : \operatorname{Ker}(f^* \otimes N) \simeq \operatorname{Ker}(\operatorname{Hom}_R(f, N)) \simeq \operatorname{Hom}_R(M, N)$, we see that the homomorphism q in (3.8) is equal to $j \otimes 1$ through this isomorphism. For it is evident from definition that $h\{(j \otimes 1)(m^* \otimes n)\}(x) = h(j(m^*) \otimes n)(x) = m^*(x)n = q(m^* \otimes n)(x)$ for $x \in M, n \in N$ and $m^* \in \operatorname{Hom}_R(M, R)$. Thus we conclude from (3.8) that $\operatorname{Tor}_1^R(\operatorname{tr}(M), N) \simeq \underline{\operatorname{Hom}}_R(M, N)$ as an R-module. Since this isomorphism is natural, it is an isomorphism of $\underline{\operatorname{End}}_R(M) \times \underline{\operatorname{End}}_R(N)$-modules. (The reader should check this.) ∎

(3.10) LEMMA. *Let M and N be CM modules over R. Suppose M is locally free on the punctured spectrum of R. Then putting $A_L = \underline{\operatorname{End}}_R(L)$ for any CM module L over R, we have the following isomorphism of $A_M \times A_N$-modules:*

$$\operatorname{Ext}_R^d(\underline{\operatorname{Hom}}_R(N, M), K_R) \simeq \operatorname{Ext}_R^1(M, (\operatorname{syz}^d \operatorname{tr}(N))'),$$

where ( )' denotes the canonical dual $\operatorname{Hom}_R(\ , K_R)$.

PROOF: There are two spectral sequences converging to the same module H_n:

$${}^1E_2^{pq} = \operatorname{Ext}_R^p(\operatorname{tr}(N), \operatorname{Ext}_R^q(M, K_R)) \Rightarrow H_n,$$

and

$${}^2E_2^{pq} = \operatorname{Ext}_R^p(\operatorname{Tor}_q^R(\operatorname{tr}(N), M), K_R) \Rightarrow H_n.$$

Here we know from the local duality theorem that $\operatorname{Ext}_R^q(M, K_R) = 0$ if $q \geq 1$. In particular we see that ${}^1E_2^{pq} = 0$ for $q \geq 1$. The first sequence ${}^1E$ is then degenerate and gives rise to an isomorphism

(3.10.1) $$H_n \simeq \operatorname{Ext}_R^n(\operatorname{tr}(N), M').$$

On the other hand, from the assumption that each localized module $M_\mathfrak{p}$ is free over $R_\mathfrak{p}$ if $\mathfrak{p}$ is a prime ideal of R with $\mathfrak{p} \neq \mathfrak{m}$, it follows that $\operatorname{Tor}_q^R(\operatorname{tr}(N), M)$ is an R-module of finite length for $q \geq 1$. Hence by the local duality theorem ${}^2E_2^{pq} = 0$ if $q \geq 1$ and if $p \neq d$. In particular we see that $H_{d+1} \simeq \operatorname{Ext}_R^d(\operatorname{Tor}_1^R(\operatorname{tr}(N), M), K_R)$. Combining this with (3.10.1) we have an isomorphism of R-modules:

$$\operatorname{Ext}_R^d(\operatorname{Tor}_1^R(\operatorname{tr}(N), M), K_R) \simeq \operatorname{Ext}_R^{d+1}(\operatorname{tr}(N), M').$$

Since this is a canonical isomorphism, it can easily be seen that it is actually an isomorphism of $A_M \times A_N$-modules. Furthermore from the definition of reduced syzygies we obtain the following isomorphisms of $A_M \times A_N$-modules:

$$\begin{aligned}
\operatorname{Ext}_R^{d+1}(\operatorname{tr}(N), M') &\simeq \operatorname{Ext}_R^d(\operatorname{syz}^1 \operatorname{tr}(N), M') \\
&\simeq \operatorname{Ext}_R^{d-1}(\operatorname{syz}^2 \operatorname{tr}(N), M') \\
&\simeq \cdots\cdots \\
&\simeq \operatorname{Ext}_R^1(\operatorname{syz}^d \operatorname{tr}(N), M'),
\end{aligned}$$

where the last module is isomorphic to $\operatorname{Ext}^1(M, (\operatorname{syz}^d \operatorname{tr}(N))')$ by local duality. Consequently we have an isomorphism of R-modules

$$\operatorname{Ext}_R^d(\operatorname{Tor}_1^R(\operatorname{tr}(N), M), K_R) \simeq \operatorname{Ext}_R^1(M, (\operatorname{syz}^d \operatorname{tr}(N))'),$$

which is natural on both variables M and N, and hence is an isomorphism of $\operatorname{End}_R(M) \times \operatorname{End}_R(N)$-modules. The lemma follows from this with (3.9). ∎

We are now ready to prove the second part of Theorem (3.4). More precisely we can show the following

(3.11) PROPOSITION. *Let M be an indecomposable CM module over R which is locally free on the punctured spectrum of R. If M is not free, then there always exists an AR sequence ending in M, and the AR translation is given by*

$$\tau(M) = (\operatorname{syz}^d \operatorname{tr}(M))'.$$

PROOF: Let $A = A_M$ be $\underline{\operatorname{End}}_R(M)$ as in (3.10) and let J be the Jacobson radical of A. First we note that A is an Artinian ring which, of course, is local. In fact we know from (3.8) that there is an exact sequence

$$\operatorname{Hom}_R(M, R) \otimes_R M \xrightarrow{q} \operatorname{End}_R(M) \longrightarrow A \longrightarrow 0.$$

Since $M_\mathfrak{p}$ is a free $R_\mathfrak{p}$-module for any prime ideal $\mathfrak{p} \neq \mathfrak{m}$, we observe that $(q)_\mathfrak{p}$ is an isomorphism for those $\mathfrak{p}$. We thus have $A_\mathfrak{p} = 0$ ($\mathfrak{p} \neq \mathfrak{m}$), which exactly means that A is an Artinian module over R. Hence it is an Artinian ring.

We have already shown that A is an Artinian local R-algebra, therefore the injective envelope $E_A(A/J)$ of A/J as a right A-module is given by $\operatorname{Ext}_R^d(A, K_R)$. Thus we obtain from (3.10) the following isomorphism of right A-modules:

$$(3.11.1) \qquad E_A(A/J) \simeq \operatorname{Ext}_R^1(M, (\operatorname{syz}^d \operatorname{tr}(M))').$$

Let s be an element of the right hand side of (3.11.1) which corresponds to the socle element of $E_A(A/J)$. If we put $\tau(M) = (\operatorname{syz}^d \operatorname{tr}(M))'$, then we know from (1.16) that $\tau(M)$ is a CM module and s gives an exact sequence of the form $0 \to \tau(M) \to E \to M \to 0$. We shall prove below that s is an AR sequence.

First assume that $\tau(M)$ is indecomposable, which will be proved in the next lemma. Then by definition s is an element of $S'(\tau(M))$. It suffices to show that s is minimal in $S'(\tau(M))$. (See (2.8)'.) To this end, let $t : 0 \to \tau(M) \to G \to L \to 0$ be another element of $S'(\tau(M))$ which satisfies $t <' s$. We want to show that $t \sim' s$. Since $t <' s$, we have an R-homomorphism $f : L \to M$ with $\operatorname{Ext}_R^1(f, \tau(M))(s) = t$. Here the homomorphism of right A-module $\operatorname{Ext}_R^1(f, \tau(M)) : \operatorname{Ext}_R^1(M, \tau(M)) \to \operatorname{Ext}_R^1(L, \tau(M))$ is a monomorphism, for it sends the socle element s to a nontrivial element t. This precisely means by (3.10) that $\operatorname{Ext}_R^d(\underline{\operatorname{Hom}}_R(M, f), K_R)$ is a monomorphism. This is equivalent by local duality to $\underline{\operatorname{Hom}}_R(M, f) : \underline{\operatorname{Hom}}_R(M, L) \to \underline{\operatorname{Hom}}_R(M, M)$ being an epimorphism of right A-modules. Thus by the definition of $\underline{\operatorname{Hom}}$, $\operatorname{Hom}_R(M, f) : \operatorname{Hom}_R(M, L) \to \operatorname{Hom}_R(M, M)$ is also an epimorphism. It then follows that f is a split morphism. Since both M and L are indecomposable, this shows that f is an isomorphism, hence $s \sim' t$.

It remains to prove that $\tau(M)$ is indecomposable. It is sufficient by the local duality theorem to show that $\operatorname{syz}^d \operatorname{tr}(M)$ is indecomposable. Thus the proposition follows from the following lemma.

(3.12) LEMMA. *Let M be an indecomposable CM module over R which is locally free on the punctured spectrum of R. Then $\operatorname{syz}^d \operatorname{tr}(M)$ is also indecomposable.*

PROOF: We divide the proof into three cases.

Case 1. $d = 0$. By definition there is an exact sequence

$$(3.12.1) \quad 0 \longrightarrow \operatorname{Hom}_R(M, R) \longrightarrow \operatorname{Hom}_R(F_0, R) \xrightarrow{f^*} \operatorname{Hom}_R(F_1, R) \longrightarrow \operatorname{tr}(M) \longrightarrow 0.$$

where $F_1 \xrightarrow{f} F_0 \to M \to 0$ is a finite free presentation of M. Recall that $\operatorname{syz}^0 \operatorname{tr}(M)$ is the module obtained from $\operatorname{tr}(M)$ by avoiding free direct summand, see definition (1.15).

Thus it is sufficient to show that $\operatorname{tr}(M)$ is indecomposable, if the presentation is minimal. Suppose not. Then, after taking suitable free bases of F_0 and F_1, the homomorphism f^* is expressed as a nontrivial direct sum of two matrices. Since the matrix of f is just a transpose of that of f^*, this shows that f is also decomposed as a direct sum of two homomorphisms. Therefore M is decomposed, which contradicts the assumption.

Case 2. $d = 1$. Write X for the image of f^* in the sequence (3.12.1). By definition, X is a direct sum of $\operatorname{syz}^1 \operatorname{tr}(M)$ with a free module. Hence it is enough to show the following: If $X = X_1 \oplus X_2$, then either X_1 or X_2 is free. For this, taking the minimal free cover $G \to \operatorname{Hom}_R(M, R) \to 0$ of $\operatorname{Hom}_R(M, R)$, we have from (3.12.1) the following exact sequence:

(3.12.2) $$G \xrightarrow{g} \operatorname{Hom}_R(F_0, R) \longrightarrow X \longrightarrow 0.$$

Then the matrix of g is in the form $\begin{pmatrix} a & 0 \\ 0 & b \end{pmatrix}$ according to the decomposition $X = X_1 \oplus X_2$. Taking the R-dual of (3.12.2), we have the sequence

$$F_1 \xrightarrow{f} F_0 \xrightarrow{g^*} \operatorname{Hom}_R(G, R),$$

which must be exact, because $\operatorname{Ext}_R^1(\operatorname{tr}(M), R) = 0$ by (3.6). Therefore we see that $M = \operatorname{Coker}(f) = \operatorname{Im}(g^*) = \operatorname{Im}(a^*) \oplus \operatorname{Im}(b^*)$, hence either $a = 0$ or $b = 0$ because of the indecomposability of M. This shows that one of X_1 and X_2 is free.

Case 3. $d \geq 2$. This case is the hardest part of the proof. First of all we observe the following:

(3.12.3) $$\operatorname{Ext}_R^n(\operatorname{Hom}_R(\operatorname{syz}^d \operatorname{tr}(M), R), R) = 0 \quad (1 \leq n \leq d - 2).$$

For the proof of this, take a free resolution of $\operatorname{tr}(M)$ by d-th term:

$$0 \longrightarrow N \longrightarrow F_{d-1} \longrightarrow \cdots \longrightarrow F_1 \longrightarrow F_0 \longrightarrow \operatorname{tr}(M) \longrightarrow 0,$$

where each F_i is a free R-module and N is a direct sum of $\operatorname{syz}^d \operatorname{tr}(M)$ with a free module. We know by (3.6) that the R-dual of this sequence $\operatorname{Hom}_R(F_0, R) \to \operatorname{Hom}_R(F_1, R) \to \cdots \to \operatorname{Hom}_R(F_{d-1}, R) \to \operatorname{Hom}_R(N, R) \to 0$ is exact. Since this gives a part of the free resolution of $\operatorname{Hom}_R(N, R)$ and since the dual of this sequence is also exact, we see that $\operatorname{Ext}_R^n(\operatorname{Hom}_R(N, R), R) = 0$ $(1 \leq n \leq d - 2)$. Thus (3.12.3) follows from this.

Now return to the proof of the lemma. Suppose $\mathrm{syz}^d\,\mathrm{tr}(M)$ is decomposed as $X \oplus Y$ with $X \neq 0, Y \neq 0$. We want to have a contradiction from this. We first note that both $\mathrm{Hom}_R(X,R)$ and $\mathrm{Hom}_R(Y,R)$ have projective dimension $\geq (d-1)$. In fact, if $\mathrm{Hom}_R(X,R)$ is free, then X is also free, because X is reflexive. In this case, $\mathrm{syz}^d\,\mathrm{tr}(M)$ contains a free summand and this contradicts the definition of reduced syzygies. Thus $\mathrm{Hom}_R(X,R)$ is nonfree, and likewise $\mathrm{Hom}_R(Y,R)$ is. If one of them has projective dimension $\leq (d-2)$, then, for some integer n ($1 \leq n \leq d-2$), we have $\mathrm{Ext}_R^n(\mathrm{Hom}_R(X,R),R) \neq 0$ or $\mathrm{Ext}_R^n(\mathrm{Hom}_R(Y,R),R) \neq 0$ and contrary to (3.12.3). Therefore both $\mathrm{Hom}_R(X,R)$ and $\mathrm{Hom}_R(Y,R)$ have projective dimension $\geq (d-1)$. In particular we see that $\mathrm{syz}^{d-2}(\mathrm{Hom}_R(X,R))$ and $\mathrm{syz}^{d-2}(\mathrm{Hom}_R(Y,R))$ are neither null nor free.

Since we know from (3.6) that $\mathrm{Ext}_R^n(\mathrm{tr}(M),R) = 0$ ($1 \leq n \leq d$), $\mathrm{Hom}_R(\mathrm{tr}(M),R)$ is isomorphic to the d-th syzygy of $\mathrm{Hom}_R(\mathrm{syz}^d\,\mathrm{tr}(M),R)$ up to free summands. (Why?) On the other hand, (3.12.1) shows that $\mathrm{Hom}_R(\mathrm{tr}(M),R)$ is the second syzygy of M. (Notice that $\mathrm{Hom}_R(\mathrm{Hom}_R(M,R),R) \simeq M$, for M is CM and locally free on the punctured spectrum of R.) Therefore M is isomorphic to the $(d-2)$-th syzygy of $\mathrm{Hom}_R(\mathrm{syz}^d\,\mathrm{tr}(M),R)$ up to free summands. As a consequence, we obtain isomorphisms

$$M \simeq \mathrm{syz}^{d-2}(\mathrm{Hom}_R(\mathrm{syz}^d\,\mathrm{tr}(M),R))$$
$$\simeq \mathrm{syz}^{d-2}(\mathrm{Hom}_R(X,R)) \oplus \mathrm{syz}^{d-2}(\mathrm{Hom}_R(Y,R)),$$

where $\simeq$ stands for the isomorphism up to free summands. As remarked before, neither module in the last term is null, so we see that M is actually decomposed. This contradiction proves the lemma. ∎

For later use we make the following remark which we have already shown in the proof of (3.11).

(3.13) REMARK. Let M be a nonfree indecomposable CM module over R that is locally free on the punctured spectrum of R, and let $\tau(M) = (\mathrm{syz}^d\,\mathrm{tr}(M))'$ as in (3.11). Then the AR sequence ending in M is a short exact sequence corresponding to the socle element in $\mathrm{Ext}_R^1(M,\tau(M))$.

Chapter 4. Auslander categories

In this chapter we will introduce Auslander's general theory by which he reached the idea of AR sequences. We do this by considering Auslander categories. The proof of Theorem (4.18) is one of our main purposes here. Theorem (4.22) is also a remarkable result due to Auslander. The theorems can be stated without using Auslander categories, but they are required in its proof.

We keep the notation of the previous chapter, so that R is a Henselian CM local ring with maximal ideal $\mathfrak{m}$ and with residue field k. The dimension of R is denoted by d. Furthermore we always assume that R has the canonical module K_R. We denote by $\mathfrak{C}(R)$, or more simply $\mathfrak{C}$, the category of all CM modules over R and R-homomorphisms. We also denote by (Ab) the category of all Abelian groups.

The idea is to move our attention from $\mathfrak{C}$ into the category of functors on $\mathfrak{C}$.

(4.1) DEFINITION. Denote by $\mathrm{Mod}(\mathfrak{C})$ the category of contravariant additive functors from $\mathfrak{C}$ to (Ab). Namely, objects in $\mathrm{Mod}(\mathfrak{C})$ are the contravariant functors $F : \mathfrak{C} \to (Ab)$ with $F(M \oplus N) = F(M) \oplus F(N)$ for any M and N in $\mathfrak{C}$ and, morphisms from F to G are the natural transformations of functors from F to G.

For a morphism $t : F \to G$ in $\mathrm{Mod}(\mathfrak{C})$ we may define the kernel and the cokernel of t as follows:

$$\mathrm{Ker}(t)(M) = \mathrm{Ker}(t(M)) \quad \text{and} \quad \mathrm{Coker}(t)(M) = \mathrm{Coker}(t(M))$$

for any object M in $\mathfrak{C}$. It is an easy exercise to show that $\mathrm{Mod}(\mathfrak{C})$ is an Abelian category.

For simplicity we write $(\ ,M)$ instead of $\mathrm{Hom}_R(\ ,M)$ for any finitely generated module M. Note that $(\ ,M)$ is in $\mathrm{Mod}(\mathfrak{C})$. Therefore we have a covariant functor $c : \mathfrak{C} \to \mathrm{Mod}(\mathfrak{C})$ by sending M to $(\ ,M)$.

(4.2) REMARK. For any $M \in \mathfrak{C}$ and $F \in \mathrm{Mod}(\mathfrak{C})$, the Abelian group $F(M)$ has the structure of an R-module in a natural way. In fact, for an element a of R, if we denote by i_a the multiplication map by a on M, then $F(i_a)$ gives an action of a on $F(M)$. More generally this shows that $F(M)$ is a right $\mathrm{End}(M)$-module. It is also easy to see that $\mathrm{Hom}_{\mathrm{Mod}(\mathfrak{C})}(F, G)$ is an R-module for $F, G \in \mathrm{Mod}(\mathfrak{C})$.

The following is known as Yoneda's lemma.

(4.3) LEMMA. *The functor* $c : \mathfrak{C} \to \mathrm{Mod}(\mathfrak{C})$ *is fully faithful.*

PROOF: It suffices to show the following is an isomorphism:

$$\varphi : (M, N) \to \mathrm{Hom}_{\mathrm{Mod}(\mathfrak{C})}((\ , M), (\ , N)),$$

where φ is given by $\varphi(f) = (\ , f)$ for any $f \in (M, N)$. If $\varphi(f) = \varphi(g)$, then we get $f = g$ by evaluating these natural transformations at R. This shows that φ is a monomorphism. Next, let $t : (\ , M) \to (\ , N)$ be any morphism in $\mathrm{Mod}(\mathfrak{C})$. Then define $g \in (M, N)$ by $g = t(M)(1_M)$. It is, then, easy to see that $\varphi(g) = t$. (Why ?) Hence φ is an epimorphism. ■

(4.4) DEFINITION. An object F in $\mathrm{Mod}(\mathfrak{C})$ is said to be **finitely generated** if there is an epimorphism $(\ , M) \to F$ with $M \in \mathfrak{C}$. And F is **finitely presented** if there is an exact sequence in $\mathrm{Mod}(\mathfrak{C})$; $(\ , N) \to (\ , M) \to F \to 0$ with M and N in $\mathfrak{C}$.

To see the following is so easy that we leave its proof to the reader as an exercise.

(4.5) *Exercise.* For finitely generated objects F and G in $\mathrm{Mod}(\mathfrak{C})$, $\mathrm{Hom}_{\mathrm{Mod}(\mathfrak{C})}(F, G)$ is a finitely generated R-module.

(4.6) DEFINITION. We denote by $\mathrm{mod}(\mathfrak{C})$ the full subcategory of $\mathrm{Mod}(\mathfrak{C})$ consisting of all finitely presented functors, and call it the **Auslander category** of $\mathfrak{C}$.

Note that the functor c defined in (4.1) makes it possible to regard $\mathfrak{C}$ as a subcategory of $\mathrm{Mod}(\mathfrak{C})$. Moreover (4.3) shows that $\mathfrak{C}$ is a full subcategory of $\mathrm{mod}(\mathfrak{C})$.

As remarked before it can be easily seen that $\mathrm{Mod}(\mathfrak{C})$ is an Abelian category. We will show in this chapter that $\mathrm{mod}(\mathfrak{C})$ is also an Abelian category. The most difficult part of this is to show that the kernel of a morphism in $\mathrm{mod}(\mathfrak{C})$ is a finitely presented functor, while the following are rather easy to prove.

(4.7) LEMMA.
(4.7.1) *If we are given an exact sequence* $F \to G \to H \to 0$ *in* $\mathrm{Mod}(\mathfrak{C})$, *and if F and G are finitely presented, then so is H.*
(4.7.2) *If we are given an exact sequence* $0 \to F \to G \to H \to 0$ *in* $\mathrm{Mod}(\mathfrak{C})$ *with F and H finitely presented, then G is also a finitely presented functor.*

PROOF: We leave the details to the reader and only outline the proof here. For (4.7.1) write finite presentations of F and G as follows: $(\ , N) \to (\ , M) \to F \to 0, (\ , P) \to$

$(\ ,Q) \to G \to 0$. Then H has the presentation of the form $(\ ,M)\oplus(\ ,P) \to (\ ,Q) \to H \to 0$.

In (4.7.2) if the presentations of F and H are given by $(\ ,N) \to (\ ,M) \to F \to 0$ and $(\ ,S) \to (\ ,T) \to H \to 0$, then the following is exact: $(\ ,N) \oplus (\ ,S) \to (\ ,M) \oplus (\ ,T) \to G \to 0$. ∎

The next proposition gives some good reasons for considering Auslander categories.

(4.8) PROPOSITION. *When we regard $\mathfrak{C}$ as a subcategory of $\mathrm{Mod}(\mathfrak{C})$ by the functor c, any objects in $\mathfrak{C}$ are projective in $\mathrm{Mod}(\mathfrak{C})$. In particular, they are also projective in $\mathrm{mod}(\mathfrak{C})$.*

PROOF: Consider a diagram in $\mathrm{Mod}(\mathfrak{C})$ with an exact row:

$$\begin{array}{ccc} F & \xrightarrow{p} & G \longrightarrow 0 \\ & & {\scriptstyle f}\uparrow \\ & & (\ ,M), \end{array}$$

where M is in $\mathfrak{C}$. We will show that there is a morphism g from $(\ ,M)$ to F so that $p \cdot g = f$. To do this, let $E = \mathrm{End}_R(M)$. Evaluating the above diagram at M, we have a diagram of right E-modules with an exact row:

$$\begin{array}{ccc} F(M) & \xrightarrow{p(M)} & G(M) \longrightarrow 0 \\ & & {\scriptstyle f(M)}\uparrow \\ & & E. \end{array}$$

Since it is evident that E itself is a projective E-module, we will have an E-module homomorphism $q: E \to F(M)$ with $f(M) = p(M) \cdot q$. Now define a natural transformation of functors $g: (\ ,M) \to F$ by $g(X)(\varphi) = F(\varphi)(q(1_M))$ for any $X \in \mathfrak{C}$ and any $\varphi \in (X,M)$. Then it is immediate that g satisfies $p \cdot g = f$, and the proposition follows. ∎

(4.9) REMARK. Let F be in $\mathrm{mod}(\mathfrak{C})$. Then by definition we have an exact sequence

(4.9.1) $\qquad\qquad (\ ,N) \xrightarrow{q} (\ ,M) \longrightarrow F \longrightarrow 0.$

Here we know from Yoneda's lemma (4.3) that q is of the form $(\ ,f)$ for some $f \in (N,M)$. Denoting by L the kernel of f, we have an exact sequence of R-modules: $0 \to L \to N \xrightarrow{f}$

M. Suppose here that $F(R) = 0$. Then f is an epimorphism and so L is also a CM module over R; (1.3.2). As a consequence we have an exact sequence in $\mathrm{mod}(\mathfrak{C})$:

(4.9.2) $\qquad 0 \longrightarrow (\ ,L) \longrightarrow (\ ,N) \longrightarrow (\ ,M) \longrightarrow F \longrightarrow 0.$

We see by (4.8) that this gives a projective resolution of F in $\mathrm{mod}(\mathfrak{C})$. Therefore we have proved that, *if $F(R) = 0$, then F has projective dimension at most 2 in* $\mathrm{mod}(\mathfrak{C})$.

(4.10) DEFINITION. An object $S(\neq 0)$ in $\mathrm{Mod}(\mathfrak{C})$ is called **simple** if it contains no nontrivial subfunctors. That is, if T is a subobject of S in $\mathrm{Mod}(\mathfrak{C})$, then we must have either $T = 0$ or $T = S$.

(4.11) *Example*. Let M be an indecomposable module in $\mathfrak{C}$. Define the functor $S_M \in \mathrm{Mod}(\mathfrak{C})$ as follows:

For an indecomposable module N in $\mathfrak{C}$, $S_M(N) = \mathrm{End}_R(N)/\mathrm{rad}(\mathrm{End}_R(N))$ if $N \simeq M$, and $S_M(N) = 0$ otherwise.

We remark that the additive functor S_M is uniquely determined by this property. Furthermore the functor is simple. For, if F is a nontrivial subfunctor of S_M, then $F(M) \neq 0$ and $F(N) = 0$ if N is indecomposable and is not isomorphic to M. Since $F(M)$ is an $\mathrm{End}_R(M)$-submodule of $S_M(M)$ by (4.2), we see that $F(M) = S_M(M)$. Thus S_M is a simple functor.

We can see that the converse is also true.

(4.12) LEMMA. *An object in* $\mathrm{mod}(\mathfrak{C})$ *is simple as an object in* $\mathrm{Mod}(\mathfrak{C})$ *if and only if it is isomorphic to S_M for some indecomposable CM module M.*

PROOF: Let $S \neq 0 \in \mathrm{mod}(\mathfrak{C})$ be a simple object in $\mathrm{Mod}(\mathfrak{C})$. Taking an indecomposable CM module M with $S(M) \neq 0$, we can see that $S(M)$ is a finitely generated $\mathrm{End}_R(M)$-module, (4.2), so that it has $S_M(M)$ as a quotient module. Then one can construct a natural transformation $\pi : S \to S_M$ such that $\pi(M)$ is an epimorphism of an $\mathrm{End}_R(M)$-module. Therefore, by the simplicity of S, $S \simeq S_M$. ∎

The following shows why we are interested in simple functors.

(4.13) PROPOSITION. *Let M be an indecomposable CM module which is not free. Then the following two conditions are equivalent.*
(4.13.1) *The functor S_M is finitely presented, i.e. $S_M \in \mathrm{mod}(\mathfrak{C})$;*
(4.13.2) *There exists an AR sequence ending in M.*

PROOF: (4.13.1) $\Rightarrow$ (4.13.2): Suppose that the finite presentation of S_M is given by $(\ ,E) \overset{(\ ,p)}{\to} (\ ,M) \to S_M \to 0$. Letting N be the kernel of p, we have the exact sequence in $\mathrm{Mod}(\mathfrak{C})$ as in a similar way to (4.9): $0 \to (\ ,N) \to (\ ,E) \to (\ ,M) \to S_M \to 0$.

Since $S_M(R) = 0$, we have an exact sequence of CM modules $0 \to N \to E \xrightarrow{p} M \to 0$, which is not split, because $S_M(M) \neq 0$. We will see that this is the AR sequence. To this end, let $f : X \to M$ be a morphism in $\mathfrak{C}$ which is not a split epimorphism. We want to show the existence of $g : X \to E$ with $p \cdot g = f$. Note that we may assume X is indecomposable. If X is not isomorphic to M, then (X,p) is an epimorphism, because $S_M(X) = 0$. So we will have an element g in (X,E) with $(X,p)(g) = f$, which clearly means $p \cdot g = f$.

Next assume $X = M$. Since f is not a split epimorphism, f belongs to $\mathrm{rad}(\mathrm{End}_R(M))$, equivalently, f goes to 0 by the natural map $(M,M) \to S_M(M)$. Since there is an exact sequence $(M,E) \to (M,M) \to S_M(M) \to 0$, this shows the existence of $g \in (M,E)$ with $p \cdot g = f$.

(4.13.2) $\Rightarrow$ (4.13.1): Let $0 \to N \to E \to M \to 0$ be an AR sequence. Define a functor S by the exact sequence $0 \to (\ ,N) \to (\ ,E) \to (\ ,M) \to S \to 0$. It suffices to show that S is equivalent to the simple functor S_M. For an indecomposable CM module M, if X is not isomorphic to M, then $S(X) = 0$, since any homomorphism $f : X \to M$ passes through $E \to M$. If $X = M$, then the image of $(M,E) \to (M,M)$ is equal to $\mathrm{rad}(\mathrm{End}_R(M))$ by (2.9). We thus have $S(M) = S_M(M)$. Since the simple functor S_M is completely characterized by these conditions, we consequently have $S = S_M$. ∎

(4.14) DEFINITION. Denote by $\underline{\mathrm{Mod}}(\mathfrak{C})$ the full subcategory of $\mathrm{Mod}(\mathfrak{C})$ whose objects are the functors F with the property $F(R) = 0$. In a similar way, we denote by $\underline{\mathrm{mod}}(\mathfrak{C})$ the full subcategory of $\mathrm{mod}(\mathfrak{C})$ consisting of functors F with $F(R) = 0$.

Let $0 \to F \to G \to H \to 0$ be an exact sequence in $\mathrm{Mod}(\mathfrak{C})$. It is immediate from the definition that if two of the objects F, G and H are in $\underline{\mathrm{Mod}}(\mathfrak{C})$, then they all belong to $\underline{\mathrm{Mod}}(\mathfrak{C})$. The following are also easy to see by (4.7).

(4.14.1) If F and H are in $\underline{\mathrm{mod}}(\mathfrak{C})$, then so is G.
(4.14.2) If F and G are in $\underline{\mathrm{mod}}(\mathfrak{C})$, then so is H.

We give some examples of objects in $\underline{\mathrm{Mod}}(\mathfrak{C})$. If M is an R-module, then $\mathrm{Ext}_R^n(\ ,M)$ $(n \geq 1)$ are clearly objects in $\underline{\mathrm{Mod}}(\mathfrak{C})$. It is also clear that the functor $\underline{\mathrm{Hom}}_R(\ ,M)$ defined in (3.7) is an object in $\underline{\mathrm{Mod}}(\mathfrak{C})$. We will see that this is actually an object in $\underline{\mathrm{mod}}(\mathfrak{C})$ if M is in $\mathfrak{C}$.

(4.15) LEMMA. *For any CM module M, the functor $\underline{\mathrm{Hom}}_R(\ ,M)$ is an object in* $\underline{\mathrm{mod}}(\mathfrak{C})$. *That is,* $\underline{\mathrm{Hom}}_R(\ ,M)$ *is a finitely presented functor.*

PROOF: Consider a free cover of M : $F \xrightarrow{q} M \to 0$. It is sufficient to show that the cokernel of $(\ ,q)$ is the functor $\underline{\mathrm{Hom}}_R(\ ,M)$. To see this, it is enough to prove that the image of $(\ ,q)$ coincides with $\mathfrak{P}(\ ,M)$. (For the definition of $\mathfrak{P}(\ ,M)$ see (3.7).) So the lemma follows from the following fact which is obvious from the definition.

(4.15.1) For a homomorphism $f \in (X, M)$ with $X \in \mathfrak{C}$, if f passes through a free module, then it passes through F. (Prove this as an exercise.) ∎

(4.16) REMARK. Let F be an object in mod($\mathfrak{C}$). By definition we have an exact sequence $(\ , E) \to (\ , M) \xrightarrow{\varphi} F \to 0$. Here note that, for any CM module X and for any $f \in \mathfrak{P}(X, M)$, $\varphi(X)(f) = 0$ in $F(X)$. For, f is decomposed as $X \to Y \to M$ for some free module Y so that $F(f)$ is a composition $F(M) \to F(Y) \to F(X)$ where $F(Y) = 0$, hence $F(f) = 0$. Since we have the commutative diagram of R-modules:

$$\begin{array}{ccc} (M, M) & \xrightarrow{\varphi(M)} & F(M) \\ {\scriptstyle (f,M)}\downarrow & & \downarrow {\scriptstyle F(f)} \\ (X, M) & \xrightarrow{\varphi(X)} & F(X), \end{array}$$

we may conclude that $\varphi(X)(f) = 0$ as stated above. Note that this means the sequence $(\ , M) \to F \to 0$ induces an exact sequence $\underline{\mathrm{Hom}}_R(\ , M) \to F \to 0$. In a same way, we will have an exact sequence

$$\underline{\mathrm{Hom}}_R(\ , E) \longrightarrow \underline{\mathrm{Hom}}_R(\ , M) \longrightarrow F \longrightarrow 0.$$

Combining this with (4.15) we see the following:

(4.17) LEMMA. *The category* mod($\mathfrak{C}$) *is an Abelian category. In particular, if* $t : F \to G$ *is a morphism in* mod($\mathfrak{C}$), *then* Ker(t) *is an object in* mod($\mathfrak{C}$).

PROOF: If we know that Ker(t) $\in$ mod($\mathfrak{C}$), then it will be easy and a good exercise to verify the rest of other properties of an Abelian category. (See (4.14.1) and (4.14.2), and remark that Mod($\mathfrak{C}$) is an Abelian category.) Here we only prove Ker(t) $\in$ mod($\mathfrak{C}$).

First we show the following as a special case of this.

(4.17.1) Let M and N be in $\mathfrak{C}$ and let t be a morphism of $\underline{\mathrm{Hom}}_R(\ , M)$ to $\underline{\mathrm{Hom}}_R(\ , N)$. Then Ker($t$) is finitely presented.

In fact, by Yoneda's lemma, t is of the form $\underline{\mathrm{Hom}}_R(\ , f)$ for some $f \in (M, N)$. Since $\underline{\mathrm{Hom}}_R(\ , X) = 0$ for a free module X, after adding a free summand to M, we may assume that f is an epimorphism of R-modules. Let L be the kernel of f and consider the exact sequence $0 \to L \xrightarrow{i} M \xrightarrow{f} N \to 0$. Note that L is a CM module. We will show that the following is exact:

(4.17.2) $\qquad\qquad \underline{\mathrm{Hom}}_R(\ , L) \xrightarrow{s} \underline{\mathrm{Hom}}_R(\ , M) \xrightarrow{t} \underline{\mathrm{Hom}}_R(\ , N)$

If this is true, then Ker(t) is finitely generated, and similarly Ker(s) is shown to be finitely generated. Hence Ker(t) is finitely presented. To see the exactness of (4.17.2), let $X \in \mathfrak{C}$ and let $g \in (X, M)$ such that the image of g in $\underline{\text{Hom}}_R(X, N)$ is 0. Then by definition we have the following commutative diagram of CM modules:

$$\begin{array}{ccccccccc} 0 & \longrightarrow & L & \xrightarrow{i} & M & \xrightarrow{f} & N & \longrightarrow & 0 \\ & & & & {\scriptstyle g}\uparrow & & {\scriptstyle b}\uparrow & & \\ & & & & X & \xrightarrow{a} & Y & & \end{array}$$

where Y is a free module. Thus there is a homomorphism $h : Y \to M$ such that $b = f \cdot h$. Since $g - h \cdot a = i \cdot k$ for some $k \in (X, L)$ and since $h \cdot a$ is 0 in $\underline{\text{Hom}}_R(X, M)$, we see that g is in the image of $s(X)$. This shows that (4.17.2) is exact.

Next consider the general case, and so let $t : F \to G$ be a morphism in $\underline{\text{mod}}(\mathfrak{C})$. We have, by (4.16), the commutative diagram with exact rows:

$$\begin{array}{ccccccccc} \underline{\text{Hom}}_R(\ , N) & \xrightarrow{a} & \underline{\text{Hom}}_R(\ , M) & \xrightarrow{b} & F & \longrightarrow & 0 \\ {\scriptstyle u}\downarrow & & {\scriptstyle v}\downarrow & & {\scriptstyle t}\downarrow & & \\ \underline{\text{Hom}}_R(\ , P) & \xrightarrow{c} & \underline{\text{Hom}}_R(\ , Q) & \xrightarrow{d} & G & \longrightarrow & 0, \end{array}$$

where N, M, P and Q are in $\mathfrak{C}$. Define then a new functor H as follows:

$$0 \longrightarrow H \xrightarrow{j} \underline{\text{Hom}}_R(\ , M) \oplus \underline{\text{Hom}}_R(\ , P) \xrightarrow{(v, -c)} \underline{\text{Hom}}_R(\ , Q).$$

It follows from (4.17.1) that H is finitely presented. We denote by q the natural projection from $\underline{\text{Hom}}_R(\ , M) \oplus \underline{\text{Hom}}_R(\ , P)$ to $\underline{\text{Hom}}_R(\ , M)$ and denote $e = b \cdot q \cdot j$. Then it is clear by chasing the diagram that e actually defines a morphism from H onto Ker(t). More precisely we have the following exact sequence:

$$\text{Ker}(c) \oplus \underline{\text{Hom}}_R(\ , N) \xrightarrow{(f, g)} H \xrightarrow{e} \text{Ker}(t) \longrightarrow 0,$$

where f is a morphism induced from the natural embedding $\text{Ker}(c) \subset \underline{\text{Hom}}_R(\ , P) \subset \underline{\text{Hom}}_R(\ , M) \oplus \underline{\text{Hom}}_R(\ , P)$ and g is a morphism induced from $a \oplus u$. Since $\text{Ker}(c) \oplus \underline{\text{Hom}}_R(\ , N)$ is finitely presented by (4.17.1) and (4.15), we see from (4.7.1) that Ker(t) is also finitely presented, which is what we wanted to prove. ∎

We are now ready to prove the main result of this chapter.

(4.18) THEOREM. (Auslander [7]) *Let M be a finitely generated R-module (not necessarily in $\mathfrak{C}$) and let n be a nonnegative integer. Then the functor $\operatorname{Ext}_R^n(\ ,M)$ is in* $\operatorname{mod}(\mathfrak{C})$, *that is, $\operatorname{Ext}_R^n(\ ,M)$ is finitely presented.*

PROOF: We prove the theorem by induction on $t = d - \operatorname{depth}(M)$.

To begin with, consider the case when $t = 0$. Since M is a CM module by definition, $(\ ,M)$ is certainly finitely presented. Denote the canonical dual $\operatorname{Hom}_R(\ ,K_R)$ by $(\ )'$ and consider a free cover of M' to obtain an exact sequence $0 \to N \to F \to M' \to 0$, where F is a free module and N is a CM module. From its dual sequence $0 \to M \to F' \to N' \to 0$, we obtain the exact sequence and isomorphisms of functors on $\mathfrak{C}$:

$$0 \longrightarrow (\ ,M) \longrightarrow (\ ,F') \longrightarrow (\ ,N') \longrightarrow \operatorname{Ext}_R^1(\ ,M) \longrightarrow 0,$$

$$\operatorname{Ext}_R^{n+1}(\ ,M) \simeq \operatorname{Ext}_R^n(\ ,N') \quad (n \geq 1).$$

(Here note that we used the fact that $\operatorname{Ext}_R^n(\ ,K_R)$, $n \geq 1$, are trivial as functors on $\mathfrak{C}$; see (1.13).) It follows from the sequence that $\operatorname{Ext}_R^1(\ ,M)$ is finitely presented. Since N' is a CM module, we see from the isomorphism that $\operatorname{Ext}_R^{n+1}(\ ,M)$ is also finitely presented, by using induction on n.

Next consider the case $t \geq 1$. Take a free cover of M to have an exact sequence $0 \to L \to G \to M \to 0$, where G is a free module and L is a module of $\operatorname{depth}(L) = \operatorname{depth}(M) + 1$. Note that we may apply the induction hypothesis to L and G. On the other hand, we have a long exact sequence:

$$0 \to (\ ,L) \xrightarrow{p_0} (\ ,G) \to (\ ,M) \to \operatorname{Ext}_R^1(\ ,L) \xrightarrow{p_1} \operatorname{Ext}_R^1(\ ,G) \to \operatorname{Ext}_R^1(\ ,M) \to \cdots$$
$$\cdots \to \operatorname{Ext}_R^n(\ ,L) \xrightarrow{p_n} \operatorname{Ext}_R^n(\ ,G) \to \operatorname{Ext}_R^n(\ ,M) \to \operatorname{Ext}_R^{n+1}(\ ,L) \xrightarrow{p_{n+1}} \cdots$$

Breaking this into short exact sequences, we have

$$0 \longrightarrow \operatorname{Coker}(p_n) \longrightarrow \operatorname{Ext}_R^n(\ ,M) \longrightarrow \operatorname{Ker}(p_{n+1}) \longrightarrow 0 \quad (n \geq 0).$$

Thus to prove that $\operatorname{Ext}_R^n(\ ,M)$, $n \geq 0$, are finitely presented, it is sufficient to show that $\operatorname{Coker}(p_n)$ and $\operatorname{Ker}(p_n)$ are finitely presented; see (4.7.2). Since $\operatorname{Ext}_R^n(\ ,L)$ and $\operatorname{Ext}_R^n(\ ,G)$ are both finitely presented by the induction hypothesis, it is immediate from (4.7.1) that $\operatorname{Coker}(p_n)$ is also finitely presented. The problem is to show that $\operatorname{Ker}(p_n)$, $n \geq 1$, is finitely presented. Since $\operatorname{Ext}_R^n(\ ,L)$ and $\operatorname{Ext}_R^n(\ ,G)$ are in $\underline{\operatorname{mod}}(\mathfrak{C})$ (c.f. (4.14)), we see from (4.17) that $\operatorname{Ker}(p_n)$ is also in $\underline{\operatorname{mod}}(\mathfrak{C})$. In particular, $\operatorname{Ker}(p_n)$ is finitely presented. This completes the proof. ∎

In a practical sense the case when $n = 1$ is of most importance in the theorem.

(4.19) COROLLARY. *The category* mod($\mathfrak{C}$) *is an Abelian category. Consequently, for any morphism* $t : F \to G$ *in* mod($\mathfrak{C}$), *the kernel of t is finitely presented.*

PROOF: If we know that Ker(t) is in mod($\mathfrak{C}$), then the other conditions for mod($\mathfrak{C}$) to be an Abelian category can easily be verified. We only prove here that Ker(t) is finitely presented. As in the proof of (4.17) it is enough to show the following:

(4.19.1) If t is a morphism of (,M) to (,N) where M and N are in $\mathfrak{C}$, then Ker(t) is finitely presented.

We know from Yoneda's lemma that t can be written as (,f) for some $f \in (M, N)$. Letting L be the kernel of f, we have an exact sequence $0 \to ($,$L) \to ($,$M) \xrightarrow{t} ($,$N)$ in Mod($\mathfrak{C}$). Note that L is not necessarily a CM module. However we know from (4.18) that Ker(t) = (,L) is finitely presented. ∎

(4.20) COROLLARY. *Let N be any finitely generated module over R (not necessarily in $\mathfrak{C}$). Then there exists a CM module M and a homomorphism $f \in (M, N)$ which satisfy the following condition:*

(4.20.1) *Any homomorphism from any CM module L into N is a composition of f with some homomorphism from L to M.*

Such a module M (or a homomorphism f) satisfying (4.20.1) is called a **CM approximation** of N.

PROOF: The functor (,N) on $\mathfrak{C}$ is, by the theorem, finitely presented. In particular, there is an epimorphism (,M) → (,N) for some CM module M. This exactly means (4.20.1). ∎

(4.21) COROLLARY. *There are only a finite number of isomorphism classes of indecomposable CM modules L from which there is an irreducible morphism into the ring R.*

PROOF: Apply (4.20) to the maximal ideal $\mathfrak{m}$ to get a CM approximation $f : M \to \mathfrak{m}$. Let $h : L \to R$ be an irreducible morphism from an indecomposable CM module L into R. Since h is not a split morphism, its image is in $\mathfrak{m}$. Thus h must be a composition $L \xrightarrow{g} M \xrightarrow{f} \mathfrak{m} \subset R$ for some $g \in (L, M)$. It then follows from the definition of irreducible morphisms that g is a split monomorphism. That is, L is isomorphic to a direct summand of M. Hence there are only a finite number of such L. ∎

We close this chapter by noting the following significant result of Auslander.

(4.22) THEOREM. (Auslander [7]) *If R has only a finite number of isomorphism classes of indecomposable CM modules, then R is an isolated singularity.*

PROOF: It will be sufficient, from (3.2), to prove that $\mathfrak{C}(R)$ admits AR sequences. To see this, let M be an indecomposable CM module which is not free. We want to show that the simple functor S_M given by M is finitely presented; cf.(4.13). To do this, let F be the kernel of the natural epimorphism $(\ ,M) \to S_M$ and let $\{N_1, N_2, \ldots, N_n\}$ be a complete list of indecomposable CM modules. Note that each $F(N_i)$ is a finitely generated R-module since it is a submodule of (N_i, M). Let $\{f_{i1}, f_{i2}, \ldots, f_{i\mu(i)}\}$ be a set of generators of $F(N_i)$. Putting $L = \sum_i N_i^{(\mu(i))}$ and considering each f_{ij} as an R-homomorphism $N_i \to M$, we define a homomorphism of R-modules $\varphi : L \to M$ by $\varphi = (f_{11}, f_{12}, \ldots, f_{n\mu(n)})$. Note that L is a CM module and we can consider a sequence in $\mathrm{Mod}(\mathfrak{C})$:

$$(\ ,L) \xrightarrow{(\ ,\varphi)} (\ ,M) \longrightarrow S_M \longrightarrow 0.$$

This is actually an exact sequence. For, substituting any N_i in this sequence, we see that $(N_i, L) \to (N_i, M) \to S_M(N_i) \to 0$ is an exact sequence of R-modules. We have thus shown that S_M is finitely presented. ∎

Chapter 5. AR quivers

Our purpose in this chapter is to give the definition of AR quivers and to discuss some general properties from that. Our main result here is Theorem (5.9) which states that if R is an isolated singularity, then the AR quiver of R is a locally finite graph. In the second half of the chapter we will give some computations to draw an AR quiver for the simplest case.

In this chapter R is a Henselian CM local ring with maximal ideal $\mathfrak{m}$ and $k = R/\mathfrak{m}$ is assumed to be an *algebraically closed field*, unless otherwise specified. We always assume that R has a canonical module. As before, $\mathfrak{C}$ denotes the category of CM modules over R and we write $(\ ,\ )$ instead of $\text{Hom}_R(\ ,\ )$.

(5.1) DEFINITION. For a homomorphism $g \in (M, N)$ with M, N CM modules, consider the following condition:

(5.1.1) Write the decompositions of M and N into indecomposable modules as $M = \sum_i M_i$ and $N = \sum_j N_j$ and decompose g along this decomposition as $g = (g_{ij})$ where $g_{ij} \in (M_i, N_j)$. Then no g_{ij} is an isomorphism.

We define a descending chain of R-submodules $\{(M, N)_n | n \geq 1\}$ of (M, N) as follows:

$$(M, N)_n = \{ f \in (M, N) \mid \text{there are } X_i \in \mathfrak{C}\ (0 \leq i \leq n, X_0 = M, X_n = N) \text{ and}$$
$$g_i \in (X_{i-1}, X_i) \text{ which satisfy the condition (5.1.1) above such that}$$
$$f = g_n \cdot g_{n-1} \cdots g_2 \cdot g_1 \}.$$

It is easy to see from the definition that $(M, N) \supseteq (M, N)_1 \supseteq (M, N)_2 \supseteq \ldots \supseteq (M, N)_n \supseteq (M, N)_{n+1} \supseteq \ldots$ is a sequence of R-submodules and that if M and N are indecomposable, then $(M, N)_1$ is the set of all nonsplit homomorphisms of M into N. In particular $(M, M)_1 = \text{rad}\,\text{End}_R(M)$ if M is indecomposable. Furthermore it is clear that

$f \in (M, N)$ is irreducible if and only if f belongs to $(M, N)_1$ but not to $(M, N)_2$; see (2.10). So it will be meaningful to define:

(5.1.2) $$\mathrm{Irr}(M, N) = (M, N)_1/(M, N)_2$$

We may say that $\mathrm{Irr}(M, N)$ is the R-module of all irreducible morphisms of M to N. Actually we see that $\mathrm{Irr}(M, N)$ is a vector space over $k = R/\mathfrak{m}$. In fact, if $f \in (M, N)_1$ and $r \in \mathfrak{m}$, then $r \cdot f$ is a composition of f with the multiplication map $M \to M$ by r, so that $r \cdot f \in (M, N)_2$. We define $\mathrm{irr}(M, N)$ as the dimension of $\mathrm{Irr}(M, N)$ as a k-vector space.

(5.1.3) $$\mathrm{irr}(M, N) = \dim_k \mathrm{Irr}(M, N)$$

Note that $\mathrm{irr}(M, N)$ is always finite, for $(M, N)_1$ is a finitely generated R-module.

Now we will define AR quivers.

(5.2) DEFINITION. The **AR quiver** Γ of R (more precisely, the AR quiver of $\mathfrak{C}$) is a graph consisting of vertices, arrows and dotted lines, where vertices are the isomorphism classes of indecomposable CM modules and we draw n arrows from $[M]$ to $[N]$ if and only if $\mathrm{irr}(M, N) = n \geq 1$. Furthermore if there is an AR sequence $0 \to \tau(M) \to E \to M \to 0$, then the vertex $[M]$ in Γ is connected by $[\tau(M)]$ with a dotted line.

(5.3) REMARK. This definition of AR quivers applies only when $k = R/\mathfrak{m}$ is algebraically closed. In general cases, the way to draw arrows in Γ should be changed as indicated now.

As remarked above $\mathrm{Irr}(M, N)$ is a vector space over k. It is precisely a right k_M-module and a left k_N-module, where $k_M = \mathrm{End}_R(M)/\mathrm{rad}(\mathrm{End}_R(M))$. Since M is indecomposable, k_M is a skew field which is an extension of k. Denote by $r_{M,N}$ (resp. $l_{M,N}$) the dimension of $\mathrm{Irr}(M, N)$ as a right k_M-(resp. left k_N-) space. Then arrows in Γ should be drawn from $[M]$ to $[N]$ with a pair of integers, $(r_{M,N}, l_{M,N})$, attached if $\mathrm{Irr}(M, N) \neq 0$.

If k is an algebraically closed field, then there will be no difference between two definitions above, because $k = k_M = k_N$.

(5.4) *Example.* Let R be a regular local ring of dimension d. Then we know from (1.5.1) that the free module R is the unique indecomposable CM module over R. It is obvious that (R, R) is isomorphic to R and under this isomorphism we easily see that $(R, R)_1 \simeq \mathfrak{m}$ and that $(R, R)_2 \simeq \mathfrak{m}^2$. In particular we have $\mathrm{Irr}(R, R) \simeq \mathfrak{m}/\mathfrak{m}^2$ and $\mathrm{irr}(R, R) = d$. Since there are no AR sequences in this case (for there are no indecomposable CM modules which are not free), the AR quiver of R consequently looks like:

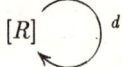

where d means that there are d distinct arrows.

Other than the above, there are no examples where we can draw the AR quiver directly from the definition. It is necessary to inquire more closely into general properties of AR quivers.

(5.5) LEMMA. *Let M and N be indecomposable CM modules over R and assume that there is an AR sequence ending in M:*

$$0 \longrightarrow \tau(M) \xrightarrow{q} E \xrightarrow{p} M \longrightarrow 0.$$

Let n be the number of copies of N in direct summands of E. Then the following equality holds:

$$\mathrm{irr}(N, M) = n.$$

PROOF: With the notation in (5.1), we denote $S(N, E) = (N, E)/(N, E)_1$. It is easy to see that $S(N, E)$ is a vector space over k. We may regard this as a space of split monomorphisms of N into E, since N is indecomposable. If X is an indecomposable CM module which is not isomorphic to N, then clearly $S(N, X) = 0$. On the other hand, $S(N, N) = \mathrm{End}_R(N)/\mathrm{rad}(\mathrm{End}_R(N)) \simeq k$, since k is algebraically closed. It thus follows that $S(N, E)$ is a vector space of dimension n. Hence it is enough to see that $S(N, E)$ is isomorphic to $\mathrm{Irr}(N, M)$ as a k-vector space. For this, define a map $\varphi : S(N, E) \to \mathrm{Irr}(N, M)$ by $\varphi(h) = p \cdot h$. We know from (2.9) that φ is a well-defined epimorphism. We want to prove it is also a monomorphism. Let h be an element of (N, E) such that $p \cdot h \in (N, M)_2$. Then by definition, there are $X \in \mathfrak{C}$ and nonsplit morphisms $a : N \to X$ and $b : X \to M$ with the commutative diagram:

$$\begin{array}{ccc} E & \xrightarrow{p} & M \\ h\uparrow & & b\uparrow \\ N & \xrightarrow{a} & X. \end{array}$$

From the property (2.9) of AR sequences, we see that there is a homomorphism $c : X \to E$ such that $b = p \cdot c$. In other words, $h - c \cdot a$ has its image in $\tau(M)$, that is, there is $g : N \to \tau(M)$ with $h - c \cdot a = q \cdot g$. Here we claim that both $c \cdot a$ and $q \cdot g$ are not split. Actually if $q \cdot g$ were split, then g would be an isomorphism hence q would be split. If $c \cdot a$ were split, then so would a, both of which are absurd. Thus we have seen that $h = c \cdot a + q \cdot g$ belongs to $(N, E)_1$, proving that φ is a monomorphism. ∎

As the dual of this we have the following lemma, which can be proved in the same way as (5.5) using (2.9)'.

(5.6) LEMMA. *Let M and N be indecomposable CM modules and assume that there is an AR sequence starting from N:*

$$0 \longrightarrow N \longrightarrow E \longrightarrow \tau^{-1}(N) \longrightarrow 0.$$

Let n' be the number of indecomposable CM modules isomorphic to M which appear in direct summands of E. Then the following equality holds:

$$\mathrm{irr}(N, M) = n'.$$

Combining (5.5) with (5.6) we have a very useful result.

(5.7) LEMMA. *Let $0 \to N \to E \to M \to 0$ be an AR sequence and let L be an indecomposable CM module. Then we have*

$$\mathrm{irr}(L, M) = \mathrm{irr}(N, L).$$

(5.8) REMARK. Let $0 \to N \to E \to M \to 0$ be an AR sequence and suppose that E is decomposed as a direct sum of indecomposable modules $\sum_i L_i^{(n_i)}$, where $L_i \not\cong L_j$ if $i \neq j$. Then the AR quiver of R is locally in the following form:

$$
\begin{array}{ccc}
 & [L_1] & \\
{}^{n_1}\nearrow & & \searrow{}^{n_1} \\
[N] \quad & \cdots & \quad [M] \\
{}_{n_r}\searrow & & \nearrow{}_{n_r} \\
 & [L_r] &
\end{array}
$$

where each n_i indicates there are n_i arrows in the same direction. This is indeed a direct consequence of (5.5), (5.6) and (5.7). In particular, this shows that if there is an AR sequence ending in M, then there are only a finite number of arrows in the AR quiver ending in $[M]$. Analogously if there is an AR sequence starting from N, then there are only a finite number of arrows from $[N]$.

From this remark together with previous results we have an important result concerning AR quivers.

(5.9) THEOREM. *Let R be an isolated singularity. Then the AR quiver Γ of R is a locally finite graph. That is, each vertex in Γ has only a finite number of arrows ending in it or starting from it.*

PROOF: Let $[M]$ be any vertex in Γ where M is an indecomposable CM module. If $M \neq R$, then by (3.2) there is an AR sequence ending in M. It thus follows from the above remark that there are only a finite number of arrows ending in $[M]$. Similarly if M is not isomorphic to the canonical module K_R, then there are only finite arrows from $[M]$. The problem is to show that the number of arrows into $[R]$ or from $[K_R]$ is finite. By the duality theorem, it is sufficient to consider the vertex $[R]$. But we proved this result in (4.21). ∎

(5.10) DEFINITION. Let $n(R)$ be the number of vertices in the AR quiver of R. In other words, $n(R)$ is the number of isomorphism classes of indecomposable CM modules over R. We say that R (or $\mathfrak{C}$) is **of finite representation type** (or **representation-finite**) if $n(R)$ is finite.

Note that a CM ring of finite representation type has only an isolated singularity; (4.22). Thus in many cases when we consider the representation type of R, we will adopt the assumption that R is an isolated singularity.

In the rest of this chapter we shall try to draw the AR quiver of a certain ring.

Let k be an algebraically closed field and let n be a positive *odd* integer. Consider the ring
$$R = k\{x,y\}/(y^2 + x^n),$$
which is called a simple curve singularity of type (A_{n-1}); see (8.5). Since n is odd, R is isomorphic to a subring $k\{t^2, t^n\}$ of the power series ring $k\{t\}$. In particular, R is an integral domain. (Note that if n is even, then R is not an integral domain.) We want to draw the AR quiver of R below. To do this, we start by determining the isomorphism classes of indecomposable CM modules over R.

First consider a subring $T = k\{x\}$ of R that is obviously a Noetherian normalization of R. Let $M (\neq 0)$ be a CM module over R. Then, by (1.9), we know that M is free as a T-module. Let μ be the maximum integer with the property $yM \subseteq x^\mu M$. We can see that $\mu < n$. In fact, if $\mu \geq n$, then $yM \subseteq x^n M = y^2 M$, hence $M = 0$. Take an element α in M with $y\alpha \in x^\mu M - x^{\mu+1} M$, and write $y\alpha = x^\mu \beta$ for some $\beta \in M$. Note that neither α nor β is in xM. Actually, if $\beta \in xM$, then $y\alpha$ would be in $x^{\mu+1}$, a contradiction. If $\alpha \in xM$, then $x^\mu \beta \in xyM \subseteq x^{\mu+1}M$, hence $\beta \in xM$, which is also a contradiction by the above. Notice also that $x^{n-\mu}M \subseteq yM \subseteq x^\mu M$. In fact, from $yM \subseteq x^\mu M$, we have $x^n M = y^2 M \subseteq yx^\mu M$, hence $x^{n-\mu} M \subseteq yM$. From the inclusion $x^{n-\mu} M \subseteq x^\mu M$, we see that $n - \mu \geq \mu$, hence the following inequality:

(i) $$0 \leq \mu \leq (n-1)/2.$$

The T-submodule N of M generated by α and β is actually an R-submodule where the action of y is defined by

(ii) $$y\alpha = x^\mu \beta, \quad y\beta = -x^{n-\mu}\alpha.$$

Secondly we show that N is a direct summand of M as a T-module. To see this, we shall prove that α and β are part of a T-free base of M. It is, thus, enough to show that β does not belong to $xM + T\alpha$. Suppose that $\beta = x\gamma + t\alpha$ ($\gamma \in M, t \in T$). Then we see from (ii) that $-x^{n-\mu}\alpha = y\beta = xy\gamma + ty\alpha = xy\gamma + tx^\mu\beta$. Hence $tx^\mu\beta = -x^{n-\mu}\alpha - xy\gamma \in x^{\mu+1}M$, since $y\gamma \in x^\mu M$ and $n - \mu > \mu$. Since x is a nonzero divisor on M, we see from this that $t\beta \in xM$. Thus $t \in xT$, for otherwise $\beta \in xM$. Consequently we have $\beta = x\gamma + t\alpha \in xM$, which is a contradiction, and so the above claim is proved.

Thirdly we prove that N is actually a direct summand of M as an R-module. To prove this, let $\{\alpha, \beta, \gamma_1, \gamma_2, \ldots, \gamma_l\}$ be a T-free base of M. Such a base exists by the previous argument. Since $yM \subseteq x^\mu M$, we have

(iii) $$y\gamma_i = x^\mu(a_i\alpha + b_i\beta + \sum_{j=1}^{l} c_{ij}\gamma_j) \quad (1 \le i \le l),$$

for some a_i, b_i and $c_{ij} \in T$. Then, letting $\gamma_i' = \gamma_i - b_i\alpha$, we see that $\{\alpha, \beta, \gamma_1', \gamma_2', \ldots, \gamma_l'\}$ is also a T-free base of M, and that

$$y\gamma_i' = x^\mu\{(a_i + \sum_{j=1}^{l} c_{ij}b_j)\alpha + \sum_{j=1}^{l} c_{ij}\gamma_j'\}.$$

Therefore, we may assume that $b_i = 0$ ($1 \le i \le l$) in (iii). Then, we have $-x^n\gamma_i = y^2\gamma_i = x^\mu(a_i y\alpha + \sum_j c_{ij}(y\gamma_j))$, hence

$$-x^{2\mu}a_i\beta - x^n\gamma_i = x^\mu \sum_{j=1}^{l} c_{ij}(y\gamma_j)$$
$$= x^\mu \sum_{j=1}^{l} c_{ij}(a_j\alpha + \sum_{k=1}^{l} c_{ik}\gamma_k)$$
$$= (x^\mu \sum_{j=1}^{l} c_{ij}a_j)\alpha + x^\mu \sum_{j,k=1}^{l} c_{ij}c_{jk}\gamma_k.$$

Since $\{\alpha, \beta, \gamma_1, \gamma_2, \ldots, \gamma_l\}$ is a free base over T, looking at the coefficient of β in the above, we get $a_i = 0$ for $1 \le i \le l$. Thus we have shown that the T-free submodule N'

generated by $\{\gamma_1, \gamma_2, \ldots, \gamma_l\}$ is closed under the action of y, hence it is an R-submodule of M. Therefore, $M = N \oplus N'$ as an R-module.

To sum up, we have shown that any indecomposable CM R-module is isomorphic to the R-module generated by α and β with the R-action defined by (ii). Note that such a module N is realized as an ideal of R generated by $\{y, x^\mu\}$, where $\alpha = x^\mu$ and $\beta = y$. We thus have proved the following:

(5.11) PROPOSITION. *Let $R = k\{x, y\}/(y^2 + x^n)$ where k is an algebraically closed field and n is an odd integer. We denote by I_μ the ideal of R generated by y and x^μ, where $I_0 = R$. Then the set $\{ I_\mu \mid 0 \leq \mu \leq (n-1)/2\}$ is a complete list of nonisomorphic indecomposable CM R-modules. In particular, the ring R is of finite representation type and $n(R) = (n+1)/2$.*

Note that in the proposition, when we regard R as $k\{t^2, t^n\}$, each module I_μ is isomorphic to a fractional ideal $(1, t^{n-2\mu})R$.

We next try to compute the AR translation. The set of non-free indecomposable CM modules is $\{I_\mu | 1 \leq \mu \leq (n-1)/2\}$ and the free resolution of such module is given by the following:

$$\cdots \xrightarrow{\rho_2} R^{(2)} \xrightarrow{\rho_1} R^{(2)} \xrightarrow{\rho_0} I_\mu \longrightarrow 0$$

where

$$\rho_2 = \begin{pmatrix} y & x^\mu \\ x^{n-\mu} & -y \end{pmatrix}, \quad \rho_1 = \begin{pmatrix} y & x^\mu \\ x^{n-\mu} & -y \end{pmatrix}, \quad \rho_0 = \begin{pmatrix} y & x^\mu \end{pmatrix}.$$

From this sequence we easily see that

(iv) $\qquad\qquad\qquad \mathrm{tr}(I_\mu) \simeq I_\mu, \quad \mathrm{syz}^1 I_\mu \simeq I_\mu \quad \text{and} \quad I_\mu^* \simeq I_\mu.$

Since R is a Gorenstein ring of dimension 1, the AR translation τ is given by $\tau(M) = (\mathrm{syz}^1 \mathrm{tr}(M))^*$; see (3.11). Hence we have

(v) $\qquad\qquad\qquad \tau(I_\mu) \simeq I_\mu \quad (1 \leq \mu \leq (n-1)/2).$

In other words, the AR sequences are of the form:

$$0 \longrightarrow I_\mu \longrightarrow E \longrightarrow I_\mu \longrightarrow 0,$$

for $1 \leq \mu \leq (n-1)/2$. Since the rank of E is two, it turns out from (5.5) and (5.6) that in the AR quiver, there are exactly two arrows starting from or ending in each I_μ.

Consequently when $1 \leq \mu \leq (n-1)/2$, we have

(vi) $$\sum_{j=0}^{(n-1)/2} \mathrm{irr}(I_j, I_\mu) = \sum_{j=0}^{(n-1)/2} \mathrm{irr}(I_\mu, I_j) = 2.$$

In the rest of this chapter we identify R with $k\{t^2, t^n\}$ and each I_μ with the fractional ideal $(1, t^{n-2\mu})R$. Clearly,

$$I_0 = R \subset I_1 \subset I_2 \subset \ldots \subset I_{(n-1)/2} = S,$$

where S is the normalization $k\{t\}$ of R. Note that for each $0 \leq i, j \leq (n-1)/2$, any R-homomorphism of I_i into I_j is of the form $u \cdot f \cdot v$ where u (resp. v) is an automorphism of I_j (resp. I_i) and f is the multiplication map of t^e for some nonnegative integer e satisfying $t^e I_i \subseteq I_j$. Note also that the set of such multiplication maps forms a semigroup given by

$$\{t^e | \ e \text{ is even with } e \geq max\{2(i-j), 0\}, \text{ or } e \text{ is odd with } e \geq n - 2j\}.$$

If $i \neq j$, then $(I_i, I_j)_1 = (I_i, I_j)$, since I_i is not isomorphic to I_j. Furthermore, if $j - i \geq 2$, then any homomorphism from I_i into I_j is a composition of two nonsplit homomorphisms $I_i \subset I_{i+1}$ and $I_{i+1} \to I_j$ by the above remarks. Hence $(I_i, I_j)_2 = (I_i, I_j)$ when $j - i \geq 2$. Similarly we see that this is true in the case when $i - j \geq 2$. Consequently we have proved

(vii) $$\mathrm{irr}(I_i, I_j) = 0 \quad \text{if} \quad |i - j| \geq 2.$$

How about the case of $j = i+1$? If $e > 0$, then the multiplication map by t^e from I_i into I_{i+1} is a composition of the natural inclusion $I_i \subset I_{i+1}$ and the multiplication map by t^e from I_{i+1} into itself. This implies that the unique candidate for the irreducible morphism from I_i to I_{i+1} is the natural inclusion $I_i \subset I_{i+1}$ up to automorphisms of I_i and I_j. Hence we have $\mathrm{irr}(I_i, I_{i+1}) \leq 1$. Analogously we see that $\mathrm{irr}(I_i, I_{i-1}) \leq 1$. To sum up,

(viii) $$\mathrm{irr}(I_i, I_j) \leq 1 \quad \text{if} \quad |i - j| = 1.$$

Next consider the case $i = j$. In this case the semigroup of multiplication by t^e is given by

$$\{t^e | \ e \text{ is even with } e \geq 0 \text{ or } e \text{ is odd with } e \geq n - 2i\}$$
$$= \{t^e | \ e = 0, 2, 4, 6, \ldots, n - 2i - 1, n - 2i, n - 2i + 1, n - 2i + 2, \ldots\}.$$

If $n - 2i \neq 1$, then it can be easily seen from this description of semigroup that any multiplication map by t^e ($e > 0$) is a composition of nonsplit morphisms, and hence $\mathrm{irr}(I_i, I_i) = 0$. On the other hand, if $n - 2i = 1$, then there is a unique candidate for the irreducible morphism, namely, multiplication by t. We thus have shown that

(ix) $\qquad \mathrm{irr}(I_i, I_i) \leq 1 \quad \text{if} \quad i = (n-1)/2,$
$\qquad \qquad \mathrm{irr}(I_i, I_i) = 0 \quad \text{if} \quad i \neq (n-1)/2.$

Taking (vi) into consideration, we easily see that all the inequalities in (vii) (viii) and (ix) must be equalities. Therefore we are able to draw the AR quiver of R:

(5.12) *The AR quiver of R is:*

$$[R] \rightleftarrows [I_1] \rightleftarrows [I_2] \rightleftarrows \cdots \rightleftarrows [I_{(n-1)/2}] \circlearrowright$$

(5.13) *Exercise.* For an *even* integer n, draw the AR quiver of $R = k\{x, y\}/(x^2 + y^n)$ where k is an algebraically closed field with odd characteristic. (See (9.9) for the answer.)

Chapter 6. The Brauer-Thrall theorem

In this chapter we shall prove one of the fundamental results concerning AR quivers, (6.2), by which we get a result analogous to the first Brauer-Thrall theorem in the representation theory of Artinian algebras.

Let k be a perfect valued field and let R be a local analytic k-algebra with maximal ideal $\mathfrak{m}$, which is assumed to be CM as before. For the definition of analytic algebras see (1.19). We denote by $\mathfrak{M}$ the category of all finitely generated R-modules and by $\mathfrak{C}$ the category of CM modules. Note that all the previous results can be applied to R, since it is a Henselian ring.

Let Γ denote the AR quiver for the category $\mathfrak{C}$. Then Γ has a numerical function on the set of vertices, called the multiplicity function:

$$e : \{\text{vertices of } \Gamma\} \longrightarrow \mathbb{N} \, ; \, [M] \longmapsto e(M)$$

where $e(M)$ is the multiplicity of M; see (1.6). We always regard the AR quiver as a graph with such a function.

(6.1) DEFINITION. A subgraph Γ° of Γ is called **of bounded multiplicity type** if there is an upper bound for the multiplicity function on the vertices in Γ°. That is, there is an integer n such that $e(M) \leq n$ for any vertex $[M]$ in Γ°.

Note that, if R is an integral domain, then this is the case only when there is a bound for the ranks of such modules, (1.6.2). Also note that, for a system of parameters $\mathbf{x} = \{x_1, x_2, ..., x_d\}$, Γ° is of bounded multiplicity type if and only if there is a bound for lengths of $M/\mathbf{x}M$ for M belonging to Γ°. It is obvious that any finite subgraph of Γ is of bounded multiplicity type. We say that Γ° is **connected** if the underlying non-oriented graph of Γ° is connected in the usual sense.

We are now ready to state the main theorem of this chapter.

(6.2) THEOREM. (Yoshino [66], Dieterich [21]) *Let Γ° be a connected component of Γ. Assume that R has only an isolated singularity and that Γ° is of bounded multiplicity type. Then $\Gamma = \Gamma^\circ$ and Γ is a finite graph. In particular, R is of finite representation type.*

If we apply this theorem to the case when Γ° is finite, then we obtain the following:

(6.3) COROLLARY. *Let R be an isolated singularity. If Γ has a finite connected component Γ°, then $\Gamma = \Gamma^\circ$ and hence R is of finite representation type.*

This shows that if $\mathfrak{C}$ is of finite representation type, then the AR quiver is a connected graph. We may apply the theorem to the infinite quiver to get:

(6.4) COROLLARY. *Let R be an isolated singularity as above. If there is a bound for multiplicities of indecomposable CM modules over R, then R is of finite representation type.*

The result corresponding to (6.4) for (noncommutative) Artinian algebras is known as the first Brauer-Thrall conjecture or Roiter-Auslander theorem, cf. Auslander [4], Ringel [55] and Roiter [56].

Our theorem is not valid unless R is an isolated singularity. We give an example now which makes Corollary (6.4) fail.

(6.5) *Example.* (Buchweitz-Greuel-Schreyer [19]) Let $R = k\{x,y\}/(x^2)$, let I_n be an ideal of R generated by $\{x, y^n\}$ for any integer $n > 0$ and set $I_0 = R$ and $I_\infty = xR$. Then $\{I_n \mid 0 \leq n \leq \infty\}$ is the complete list of non-isomorphic indecomposable CM modules over R. In particular $\mathfrak{C}$ is not of finite representation type. On the other hand, $e(I_n) = 2$ for any n ($0 \leq n < \infty$) and $e(I_\infty) = 1$.

PROOF: Let $T = k\{y\}$. Then R is a finite T-algebra and any CM module is free over T. Thus giving a CM module M over R is equivalent to giving a T-algebra map f_M from R into a matrix algebra over T and this is also equivalent to giving a square matrix A_M with entries in T and with $A_M^2 = 0$ (by taking $f_M(x) = A_M$). It can be seen that two CM modules M and N are R-isomorphic if and only if $A_M = PA_NP^{-1}$ for some invertible matrix P with entries in T. Thus the classification of all CM modules over R is the same as the classification of square-zero matrices over T up to equivalence. Since T is a discrete valuation ring, it is easily seen that any square-zero matrix is equivalent to a direct sum of matrices of the following forms:

$$(0), \begin{pmatrix} 0 & 1 \\ 0 & 0 \end{pmatrix}, \text{ or } \begin{pmatrix} 0 & y^n \\ 0 & 0 \end{pmatrix} \text{ for some integer } n > 0.$$

These matrices correspond to I_∞, I_0 and I_n respectively. ∎

The rest of this chapter is devoted to the proof of Theorem (6.2). We need some preliminaries.

As before R is a local analytic CM algebra over a perfect field k. Taking a system of parameters $\mathbf{x} = \{x_1, x_2, \ldots, x_d\}$ for R, we can form a convergent power series ring $T = k\{x_1, x_2, \ldots, x_d\}$ where d is the dimension of R. Note that R is always module-finite over T and hence it is T-free, because R is CM.

(6.6) DEFINITION. Let $R^e = R \otimes_T R$ and call it the enveloping algebra of R over T. Let $\mu : R^e \to R$ be the multiplication mapping. The **Noetherian different** $\mathcal{N}_T^R$ of R over T is defined as follows:

$$\mathcal{N}_T^R = \mu(\mathrm{Ann}_{R^e}(\mathrm{Ker}(\mu))).$$

If R is reduced (and CM) then $\mathcal{N}_T^R$ is known to coincides with the Dedekind different $\mathcal{D}_T^R$ which is given by

$$\mathcal{D}_T^R = \{f(\mathrm{trace}) \in R \mid f \in \mathrm{Hom}_R(\mathrm{Hom}_T(R,T), R)\},$$

where trace denotes the trace map of the total quotient ring $Q(R)$ of R over the quotient field $Q(T)$ of T. This also equals the inverse ideal of $\mathcal{C}_T^R = \{x \in Q(R) \mid \mathrm{trace}(xR) \subset T\}$. For more details see Scheja-Storch [57].

Recall that a system of parameters $\mathbf{x}$ is separable if the extension $Q(R)/Q(T)$ is separable; see (1.20). If this is the case, then it is known that $\mathcal{D}_T^R$ is an ideal of pure height 1 (purity of branch locus). Thus we have:

(6.7) LEMMA. *If R is a local analytic CM k-algebra which is reduced and if $\mathbf{x} = \{x_1, x_2, \ldots, x_d\}$ is a separable system of parameters for R, then $\mathcal{N}_T^R$ is an ideal of pure height one where $T = k\{x_1, x_2, \ldots, x_d\}$.*

We also note here that there always exists a separable system of parameters for a reduced analytic algebra R, since k is perfect; see (1.20).

The following is one of the most important properties of Noetherian differents.

(6.8) LEMMA. *Let R be a T-algebra as above and let M be any R^e-module, i.e. an R-bimodule with T acting centrally. Then $\mathcal{N}_T^R$ annihilates the i-th Hochschild cohomology $H_T^i(R, M)$ for $i > 0$.* (For the Hochschild cohomology see Hochschild [39] or Pierce [51].)

Since it is known that $H_T^i(R, \ ) = 0$ ($i \geq 1$) only when R is a separable algebra over T, as a special case of the lemma, we have:

(6.9) COROLLARY. *The Noetherian different $\mathcal{N}_T^R$ defines the ramification locus of* Spec(R) *over* Spec(T). *That is, for a prime ideal $\mathfrak{P}$ of R, $R_\mathfrak{P}$ is ramified over $T_{\mathfrak{P} \cap T}$ if and only if $\mathfrak{P}$ contains $\mathcal{N}_T^R$.*

From these observations we have the following:

(6.10) LEMMA. *Let $\{\mathfrak{P}_1, \mathfrak{P}_2, \ldots, \mathfrak{P}_n\}$ be a set of prime ideals of R with the property that each $R_{\mathfrak{P}_i}$ ($i = 1, 2, \ldots, n$) is a regular local ring of the same dimension t. Then there is a system of parameters $\{x_1, x_2, \ldots, x_d\}$ for R such that $\mathcal{N}_{k\{x_1, x_2, \ldots, x_d\}}^R$ is not contained in any $\mathfrak{P}_i$ ($i = 1, 2, \ldots, n$).*

PROOF: Let S denote the set $R - \cup_i \mathfrak{P}_i$ which is multiplicatively closed in R. Note that $S^{-1}R$ is a semi-local ring with maximal ideals $\mathfrak{P}_i(S^{-1}R)$ ($1 \leq i \leq n$). Since $R_{\mathfrak{P}_i} = (S^{-1}R)_{\mathfrak{P}_i(S^{-1}R)}$ is a regular local ring of dimension t, one can choose a set of elements $\{x_1, x_2, \ldots, x_t\}$ in R which forms a regular system of parameters for each $R_{\mathfrak{P}_i}$ ($1 \leq i \leq n$). Thus $(x_1, x_2, \ldots, x_t)R_{\mathfrak{P}_i} = \mathfrak{P}_i R_{\mathfrak{P}_i}$ for any i. Since $R/\sqrt{(x_1, x_2, \ldots, x_t)R}$ is a reduced analytic algebra over k and since k is perfect, we can take $\{x_{t+1}, x_{t+2}, \ldots, x_d\}$ from R which forms a separable system of parameters for $R/\sqrt{(x_1, x_2, \ldots, x_t)R}$; (1.20). Then, since $\{x_1, x_2, \ldots, x_d\}$ is a system of parameters for R, we can consider the power series ring $T = k\{x_1, x_2, \ldots, x_d\}$ on which R is module-finite. Note that, for any i ($1 \leq i \leq n$), $\mathfrak{P}_i \cap T = (x_1, x_2, \ldots, x_t)T$ which we denote by $\mathfrak{p}$. Then it is easy to see that each $R_{\mathfrak{P}_i}$ ($1 \leq i \leq n$) is unramified over $T_\mathfrak{p}$, for $R_{\mathfrak{P}_i}/\mathfrak{p} R_{\mathfrak{P}_i} = (R/\sqrt{(x_1, x_2, \ldots, x_t)R})_{\mathfrak{P}_i}$ is separable over $T_\mathfrak{p}/\mathfrak{p} T_\mathfrak{p}$. We thus conclude from (6.9) that $\mathfrak{P}_i$ does not contain $\mathcal{N}_T^R$. ∎

Now we define an ideal of R which seems to be a good invariant of an analytic algebra R.

(6.11) DEFINITION.
$$\mathcal{N}^R = \sum \mathcal{N}_{k\{x_1, x_2, \ldots, x_d\}}^R$$
where $\{x_1, x_2, \ldots, x_d\}$ ranges through all systems of parameters for R.

As a corollary of (6.10) we see that $\mathcal{N}^R$ has the following property.

(6.12) LEMMA. *The ideal $\mathcal{N}^R$ defines the singular locus of* Spec(R), *that is, for a prime ideal $\mathfrak{P}$ of R, $R_\mathfrak{P}$ is regular if and only if $\mathfrak{P}$ does not contain $\mathcal{N}^R$.*

PROOF: If $\mathfrak{P}$ does not contain $\mathcal{N}^R$, then it does not contain $\mathcal{N}_T^R$ for some $T = k\{x_1, x_2, \ldots, x_d\}$. Thus $R_\mathfrak{P}$ is unramified over $T_{\mathfrak{P} \cap T}$ by (6.9), which implies that $R_\mathfrak{P}$ is regular. On the other hand, assume that $R_\mathfrak{P}$ is regular. Then by (6.10) $\mathfrak{P}$ does not contain $\mathcal{N}_T^R$ for some $T = k\{x_1, x_2, \ldots, x_d\}$, in particular, it does not contain $\mathcal{N}^R$. ∎

(6.13) REMARK. If $R = T[x]/(f(x))$, then it is known that $\mathcal{N}_T^R$ is an ideal generated by the derivative $f'(x)$. Thus in the case when R is a hypersurface $k\{x_1, x_2, \ldots, x_{d+1}\}/(f)$,

$\mathcal{N}^R$ is generated by derivatives $\partial f/\partial x_i$ ($1 \le i \le d+1$). By this fact we are able to see that the assumption that k is a perfect field is indispensable in (6.12). For example, if there is an element a in k which is not in k^p, where p is the characteristic of k, then consider a k-algebra $R = k\{x,y\}/(x^p + ay^p)$. In this example, R is an integral domain of dimension 1, hence an isolated singularity, though $\mathcal{N}^R = 0$.

(6.14) LEMMA. *If R has only an isolated singularity, then one can choose a system of parameters* $\mathbf{x} = \{x_1, x_2, \ldots, x_d\}$ *for R which satisfies the condition:*

(6.14.1) *for any i ($1 \le i \le d$) there is a regular subring T_i of R on which R is finite and x_i belongs to the Noetherian different $\mathcal{N}_{T_i}^R$.*

We call such an $\mathbf{x}$ an **efficient system of parameters for** R.

PROOF: By induction on j ($1 \le j \le d$), one can choose a part $\{x_1, x_2, \ldots, x_j\}$ of the system of parameters for R, such that x_i belongs to $\mathcal{N}_{T_i}^R$ for some T_i ($1 \le i \le j$). This is obvious if $j = 1$, since $\mathcal{N}_T^R$ is an ideal of pure height 1 for any regular subring T of R. For $2 \le j \le d$, assume that $\{x_1, x_2, \ldots, x_{j-1}\}$ are already chosen. Then by (6.10) there is some T_j with the property that $\mathcal{N}_{T_j}^R$ is not contained in any minimal prime ideals of $(x_1, x_2, \ldots, x_{j-1})R$. Thus there is an element x_j in $\mathcal{N}_{T_j}^R$ such that $\{x_1, x_2, \ldots, x_j\}$ forms a subsystem of parameters for R. ∎

Efficient systems of parameters will play a central role in the rest of this chapter. The following proposition will be a key for using these parameters. In fact, it leads Propositions (6.16) to (6.18), which, we should mention, are generalizations of Marranda's theorem about finite group rings over complete discrete valuation rings.

(6.15) PROPOSITION. *Let $\mathbf{x} = \{x_1, x_2, \ldots, x_d\}$ be an efficient system of parameters for R and let M, N be CM modules over R. We denote by $\mathbf{x}^{(n)}$ the ideal of R generated by $\{x_1^n, x_2^n, \ldots, x_d^n\}$. Then for any R-homomorphism φ from $M/\mathbf{x}^{(2)}M$ to $N/\mathbf{x}^{(2)}N$, there exists an R-homomorphism ψ from M to N such that $\varphi \otimes R/\mathbf{x}R = \psi \otimes R/\mathbf{x}R$.*

PROOF: By induction on $d - i$ ($0 \le i \le d$) we prove the following:

(6.15.i) There is an R-homomorphism φ_i from $M/(x_1^2, \ldots, x_i^2)M$ to $N/(x_1^2, \ldots, x_i^2)N$ such that $\varphi_i \otimes R/\mathbf{x}R = \varphi \otimes R/\mathbf{x}R$.

There is nothing to prove for $i = d$, for $\varphi_d = \varphi$ is given. Assume that φ_{i+1} is already constructed ($0 \le i \le d-1$). It is enough to show the existence of φ_i from $M/(x_1^2, \ldots, x_i^2)M$ to $N/(x_1^2, \ldots, x_i^2)N$ satisfying $\varphi_i \otimes R/(x_1^2, \ldots, x_i^2, x_{i+1})R = \varphi_{i+1} \otimes R/(x_1^2, \ldots, x_i^2, x_{i+1})$. For simplicity we denote the ideal generated by $\{x_1^2, \ldots, x_i^2\}$ (resp. $\{x_1^2, \ldots, x_i^2, x_{i+1}\}$) by $\mathbf{y}_i$ (resp. $\mathbf{z}_i$). Since N is a CM module over R, we have the commutative diagram with exact rows:

$$\begin{array}{ccccccccc}
0 & \longrightarrow & N/\mathbf{y}_i N & \xrightarrow{x_{i+1}^2} & N/\mathbf{y}_i N & \longrightarrow & N/\mathbf{y}_{i+1} N & \longrightarrow & 0 \\
& & \downarrow{\scriptstyle x_{i+1}} & & \| & & \downarrow & & \\
0 & \longrightarrow & N/\mathbf{y}_i N & \xrightarrow{x_{i+1}} & N/\mathbf{y}_i N & \longrightarrow & N/\mathbf{z}_i N & \longrightarrow & 0.
\end{array}$$

Applying the functor $\mathrm{Hom}_{T_{i+1}}(M,\ )$ to this diagram where T_{i+1} being as in (6.14.1), we obtain the following commutative diagram:

$$\begin{array}{ccccccc}
0 \to \mathrm{Hom}_{T_{i+1}}(M, N/\mathbf{y}_i N) & \to & \mathrm{Hom}_{T_{i+1}}(M, N/\mathbf{y}_i N) & \to & \mathrm{Hom}_{T_{i+1}}(M, N/\mathbf{y}_i N) & \to & 0 \\
\downarrow{\scriptstyle x_{i+1}} & & \| & & \downarrow & & \\
0 \to \mathrm{Hom}_{T_{i+1}}(M, N/\mathbf{y}_i N) & \to & \mathrm{Hom}_{T_{i+1}}(M, N/\mathbf{y}_i N) & \to & \mathrm{Hom}_{T_{i+1}}(M, N/\mathbf{z}_i N) & \to & 0,
\end{array}$$

where the rows are exact, since M is a free T_{i+1}-module. Note that these rows are also exact sequences of R-bimodules. (The left (resp. right) action of R on $\mathrm{Hom}(M, N')$ is given by the one induced from the action on N' (resp. M).) Noting that $H^0_{T_{i+1}}(R, \mathrm{Hom}_{T_{i+1}}(M, N')) = \mathrm{Hom}_R(M, N')$ for any R-modules M and N', we now get the commutative diagram with exact rows by taking the Hochschild cohomology functor:

$$\begin{array}{ccccc}
\mathrm{Hom}_R(M, N/\mathbf{y}_i N) & \to & \mathrm{Hom}_R(M, N/\mathbf{y}_{i+1} N) & \to & H^1_{T_{i+1}}(R, \mathrm{Hom}_{T_{i+1}}(M, N/\mathbf{y}_i N)) \\
\| & & \downarrow & & \downarrow{\scriptstyle x_{i+1}} \\
\mathrm{Hom}_R(M, N/\mathbf{y}_i N) & \to & \mathrm{Hom}_R(M, N/\mathbf{z}_i N) & \to & H^1_{T_{i+1}}(R, \mathrm{Hom}_{T_{i+1}}(M, N/\mathbf{y}_i N)).
\end{array}$$

By (6.8) and by our choice of x_{i+1} and T_{i+1} in (6.14.1) we know that x_{i+1} on the right vertical arrow induces the trivial map. Therefore some easy diagram chasing shows that for any φ_{i+1} in $\mathrm{Hom}_R(M, N/\mathbf{y}_{i+1} N)$, there is φ_i in $\mathrm{Hom}_R(M, N/\mathbf{y}_i N)$ with $\varphi_i \otimes R/\mathbf{z}_i R = \varphi_{i+1} \otimes R/\mathbf{z}_i R$. This completes the proof of the proposition. ∎

As a direct consequence of (6.15) we obtain the following:

(6.16) PROPOSITION. *Let $\mathbf{x} = \{x_1, x_2, \ldots, x_d\}$ be an efficient system of parameters for R and let M be a CM module over R. Then M is an indecomposable R-module if and only if $M/\mathbf{x}^{(2)} M$ is indecomposable.*

PROOF: If M is decomposable, then it is obviously true that $M/\mathbf{x}^{(2)} M$ is also decomposable. Conversely assume that M is indecomposable. Take an idempotent e in $\mathrm{End}_R(M/\mathbf{x}^{(2)} M)$. We want to prove that either $e = 1$ or 0. We have a commutative diagram of natural ring homomorphisms:

$$\begin{array}{ccc}
\mathrm{End}_R(M) & \xrightarrow{\otimes R/\mathbf{x}^{(2)} R} & \mathrm{End}_R(M/\mathbf{x}^{(2)} M) \\
{\scriptstyle \alpha = \otimes R/\mathbf{x} R} \searrow & & \swarrow {\scriptstyle \beta = \otimes R/\mathbf{x} R} \\
& \mathrm{End}_R(M/\mathbf{x} M) &
\end{array}$$

Now we denote by A the image of α which is also local, for it is a homomorphic image of the local algebra $\mathrm{End}_R(M)$. It then follows from (6.15) that $\beta(e)$ belongs to A. Since $e^2 = e$, $\beta(e)$ is also an idempotent of A, hence either $\beta(e) = 1$ or 0. If $\beta(e) = 0$, then $e(M/\mathbf{x}^{(2)}M) \subset \mathbf{x}(M/\mathbf{x}^{(2)}M)$ hence $e = 0$, because $e^2 = e$. If $\beta(e) = 1$, then $\beta(1-e) = 0$ hence by the above $e = 1$. ∎

We also obtain from (6.15) the following:

(6.17) PROPOSITION. *Let $\mathbf{x} = \{x_1, x_2, \ldots, x_d\}$ be an efficient system of parameters and let $s : 0 \to N \xrightarrow{q} E \xrightarrow{p} M \to 0$ be an exact sequence in the category $\mathfrak{C}$. Denote by $\tilde{s}$ the sequence obtained from s by tensoring $R/\mathbf{x}^{(2)}R$:*

$$\tilde{s} : 0 \longrightarrow N/\mathbf{x}^{(2)}N \xrightarrow{\tilde{q}} E/\mathbf{x}^{(2)}E \xrightarrow{\tilde{p}} M/\mathbf{x}^{(2)}M \longrightarrow 0$$

(Note that $\tilde{s}$ is also exact, since $\mathbf{x}^{(2)}$ is a regular sequence on M.) If $\tilde{s}$ is split, then so is s.

PROOF: Assume $\tilde{s}$ is split, that is, there is an f in $\mathrm{Hom}_R(M/\mathbf{x}^{(2)}M, E/\mathbf{x}^{(2)}E)$ such that $\tilde{p} \cdot f$ is the identity on $M/\mathbf{x}^{(2)}M$. Then (6.15) shows that there is a g in $\mathrm{Hom}_R(M, E)$ such that $g \otimes R/\mathbf{x}R = f \otimes R/\mathbf{x}R$. Thus $(p \cdot g) \otimes R/\mathbf{x}R = (p \otimes R/\mathbf{x}R) \cdot (g \otimes R/\mathbf{x}R)$ is the identity mapping on $M/\mathbf{x}M$. In particular we see by Nakayama's lemma that $p \cdot g$ is an epimorphism, and so it must be an automorphism on M. This shows s is a split sequence. ∎

The next proposition will be useful later.

(6.18) PROPOSITION. *Let $\mathbf{x} = \{x_1, x_2, \ldots, x_d\}$ be an efficient system of parameters and let M, N be indecomposable CM modules over R. If $M/\mathbf{x}^{(2)}M$ is isomorphic to $N/\mathbf{x}^{(2)}N$, then M is isomorphic to N.*

PROOF: Let $\tilde{f}$ be an isomorphism from $M/\mathbf{x}^{(2)}M$ onto $N/\mathbf{x}^{(2)}N$. Then by (6.15) we have a homomorphism f from M to N such that $f \otimes R/\mathbf{x}R = \tilde{f} \otimes R/\mathbf{x}R$. In particular f is epimorphic by Nakayama's lemma. Thus we obtain an exact sequence $0 \to \mathrm{Ker}(f) \to M \xrightarrow{f} N \to 0$, where it is easily seen that $\mathrm{Ker}(f)$ is also a CM module. Tensoring $R/\mathbf{x}R$ with this sequence we have an exact sequence:

$$0 \longrightarrow \mathrm{Ker}(f) \otimes R/\mathbf{x}R \longrightarrow M \otimes R/\mathbf{x}R \xrightarrow{f \otimes R/\mathbf{x}R} N \otimes R/\mathbf{x}R \longrightarrow 0.$$

Since $f \otimes R/\mathbf{x}R$ is an isomorphism, we see that $\mathrm{Ker}(f) \otimes R/\mathbf{x}R = 0$, hence $\mathrm{Ker}(f) = 0$ again by Nakayama's lemma. Thus f gives an isomorphism between M and N. ∎

(6.19) REMARK. The above proof also gives the following:

If f is an R-homomorphism between CM modules M and N, and if $f \otimes R/\mathbf{x}^{(2)}R$ gives an isomorphism, then f is also an isomorphism.

For the proof of Theorem (6.2) we need the following lemma.

(6.20) LEMMA. (Harada-Sai lemma for CM modules) *Let M_i ($0 \leq i \leq 2^n$) be indecomposable CM modules over R and let $\mathbf{x} = \{x_1, x_2, \ldots, x_d\}$ be an efficient system of parameters. Let $f_i : M_{i-1} \to M_i$ ($1 \leq i \leq 2^n$) be non-isomorphic homomorphisms. Assume that $length(M_i/\mathbf{x}^{(2)}M_i) \leq n$ for $0 \leq i \leq 2^n$. Then we have*

$$(f_{2^n} \cdots f_2 \cdot f_1) \otimes R/\mathbf{x}^{(2)}R = 0.$$

PROOF: We denote $M/\mathbf{x}^{(2)}M$, $f \otimes R/\mathbf{x}^{(2)}R$ respectively by $\tilde{M}$, $\tilde{f}$. Then by (6.16) and (6.19), the $\tilde{R}$-modules $\tilde{M}_i$ and $\tilde{R}$-homomorphisms $\tilde{f}_i$ satisfy the following conditions:
(a) $\tilde{M}_i$ ($0 \leq i \leq 2^n$) are indecomposable,
(b) $length(\tilde{M}_i) \leq n$ for all i, and
(c) $\tilde{f}_i$ ($1 \leq i \leq 2^n$) are all non-isomorphic.
Then Harada-Sai lemma ([51, Proposition 7.2] or [34]) shows that the composition $\tilde{f}_{2^n} \ldots \tilde{f}_2 \cdot \tilde{f}_1$ is trivial. ∎

In the case when we are given a sequence of irreducible morphisms:

(*) $\qquad\qquad M_0 \xrightarrow{f_1} M_1 \xrightarrow{f_2} M_2 \longrightarrow \cdots \xrightarrow{f_n} M_n$

with all M_i indecomposable CM modules, then we will call this a **chain of irreducible morphisms of length** n. A chain (*) of irreducible morphisms is said to be **nontrivial with respect to a system of parameters** $\mathbf{x} = \{x_1, x_2, \ldots, x_d\}$ provided that $(f_n \cdot f_{n-1} \cdots f_1) \otimes R/\mathbf{x}R$ is a nontrivial homomorphism. The following is the result corresponding to [55, Lemma 2.1].

(6.21) LEMMA. *Let R be an isolated singularity and let $\mathbf{x} = \{x_1, x_2, \ldots, x_d\}$ be an efficient system of parameters for R. Let M, N be indecomposable CM modules over R. Assume that there is a morphism h from M into N satisfying $h \otimes R/\mathbf{x}^{(2)}R \neq 0$, and that there exists no chain of irreducible morphisms from M to N of length $< n$ which is nontrivial with respect to $\mathbf{x}^{(2)}$. Then*
(a) *there exists a chain of irreducible morphisms:*

$$M = M_0 \xrightarrow{f_1} M_1 \xrightarrow{f_2} M_2 \longrightarrow \cdots \longrightarrow M_{n-1} \xrightarrow{f_n} M_n$$

and a morphism $g : M_n \to N$ with $(g \cdot f_n \cdot f_{n-1} \cdots f_1) \otimes R/\mathbf{x}^{(2)}R \neq 0$; and

(b) *there exists a chain of irreducible morphisms:*

$$N_n \xrightarrow{g_n} N_{n-1} \xrightarrow{g_{n-1}} N_{n-2} \longrightarrow \cdots \longrightarrow N_1 \xrightarrow{g_1} N_0 = N$$

and a morphism $f : M \to N_n$ *with* $(g_1 \cdot g_2 \cdots g_n \cdot f) \otimes R/\mathbf{x}^{(2)}R \neq 0$.

PROOF: We only prove (b), for (a) can be obtained by a dual argument. The proof proceeds by induction on n. For $n = 0$, there is nothing to prove. Assume $n > 0$. Then by the induction hypothesis we have irreducible morphisms $g_i : N_i \to N_{i-1}$ ($1 \leq i \leq n-1$) with $N_0 = N$ and a morphism $f : M \to N_{n-1}$ such that $(g_1 \cdot g_2 \cdots g_{n-1} \cdot f) \otimes R/\mathbf{x}^{(2)}R \neq 0$. Our assumption implies that f can never be an isomorphism. We consider two cases: First let N_{n-1} be isomorphic to R. In this case by (4.20) we have a CM module L and a morphism h from L to N_{n-1} such that f can be factored through h:

$$\begin{array}{ccc} M & \xrightarrow{f} & N_{n-1} = R \\ {}_{h'}\searrow & & \nearrow_{h} \\ & L & \end{array}$$

We decompose L into a direct sum of indecomposable CM modules L_i, and also decompose h into a direct sum of h_i, and h' into a sum of h'_i. Therefore $f = \sum h_i \cdot h'_i$. Note that, if we take $(\ , L) \to (\ , \mathfrak{m})$ to be minimal, then the h_i are irreducible. Now, since $(g_1 \cdot g_2 \cdots g_{n-1} \cdot f) \otimes R/\mathbf{x}^{(2)}R \neq 0$, it follows that $(g_1 \cdot g_2 \cdots g_{n-1} \cdot h_i \cdot h'_i) \otimes R/\mathbf{x}^{(2)}R \neq 0$ for some i. Letting $N_n = L_i$ and $g_n = h_i$, the lemma follows in this case.

In the second case assume that N_{n-1} is not isomorphic to R. Thus there exists an AR sequence $0 \to \tau(N_{n-1}) \to L \xrightarrow{h} N_{n-1} \to 0$; see (3.2). Let $L = \sum_i L_i$ with L_i indecomposable, and let $h = (h_i)$. Again the h_i are irreducible. By the property of AR sequences one can lift f to L, thus f will be factored again in the form $f = \sum h_i \cdot h'_i$. In the same way as above, one obtains $(g_1 \cdots g_{n-1} \cdot h_i \cdot h'_i) \otimes R/\mathbf{x}^{(2)}R \neq 0$ for some i, and the proof is completed. ∎

Now we proceed to the proof of Theorem (6.2). Let R be an isolated singularity as in the theorem and let Γ° be a connected component of the AR quiver Γ. Assume that all indecomposable CM modules in Γ° are of multiplicity $\leq a$. Let $\mathbf{x} = \{x_1, x_2, \ldots, x_d\}$ be an efficient system of parameters for R. Note that for a module in Γ° we have $length(M/\mathbf{x}^{(2)}M) \leq m$ where $m = a \cdot b^d$ with b being the least integer satisfying $\mathfrak{m}^b \subset \mathbf{x}^{(2)}R$; (1.7).

Let M, N be two indecomposable CM modules with the property that there is a morphism f from M to N such that $f \otimes R/\mathbf{x}^{(2)}R \neq 0$. Assume that M belongs to Γ°. We want to prove that there is a chain of irreducible morphisms from M to N of length

$< 2^m$ which is nontrivial with respect to $\mathbf{x}^{(2)}$, and thus that N is also in Γ°. For otherwise, by (6.21) there is a chain of irreducible morphisms:

$$M = M_0 \xrightarrow{f_1} M_1 \xrightarrow{f_2} M_2 \longrightarrow \cdots \longrightarrow M_{n-1} \xrightarrow{f_n} M_n$$

and a morphism $g : M_n \to N$ with $(g \cdot f_n \cdots f_2 \cdot f_1) \otimes R/\mathbf{x}^{(2)}R \neq 0$, where $n = 2^m$. Here we note that M_i are all in Γ°, for M_i being connected with M in Γ. In particular, we have $length(M/\mathbf{x}^{(2)}M) \leq m$ by the assumption. Then (6.20) shows that $(f_n \cdot f_{n-1} \cdots f_1) \otimes R/\mathbf{x}^{(2)}R = 0$, which is a contradiction.

Summarizing the above, we have obtained the following:

(6.22.1) Let M and N be indecomposable CM modules with the property that there is a homomorphism f from M to N satisfying $f \otimes R/\mathbf{x}^{(2)}R \neq 0$. If M belongs to Γ°, then there is a chain of irreducible morphisms from M to N of length $< n(= 2^m)$. In particular N also belongs to Γ°.

The dual argument gives the dual statement of the above:

(6.22.2) Let M and N be as in (6.22.1). If N belongs to Γ°, then there is a chain of irreducible morphisms from M to N of length $< n$. In particular M also belongs to Γ°.

Now let M be any indecomposable CM module over R. Then there is a map $f : R \to M$ with $f \otimes R/\mathbf{x}^{(2)}R \neq 0$. (It is enough to take an element x in M which is not in $\mathbf{x}^{(2)}M$ and to define $f(r) = r \cdot x$.) Taking M from the vertices in Γ°, one can show by (6.22.2) that R belongs to Γ°. It then follows from (6.22.1) that any M from Γ belongs to Γ°. Thus we have proved that $\Gamma = \Gamma^\circ$, and at the same time, that any vertex in Γ is connected with R by a directed path of length at most n. On the other hand we know from (5.9) that the graph Γ is locally finite. Hence Γ must be a finite graph and the proof is finished. ∎

Chapter 7. Matrix factorizations

In this chapter we shall make a brief presentation of Eisenbud's matrix factorization theorem, (7.4). This will play a key role when we treat CM modules over a hypersurface singularity.

Suppose that a ring R is a homomorphic image of a regular local ring S, that is, $R = S/I$ for some ideal I of S. When I can be chosen as a principal ideal (f), we call R a **local ring of hypersurface** (or simply a **hypersurface**) defined by f in S. Throughout this chapter R is always a hypersurface and is always given by $R = S/(f)$, which is assumed to be Henselian. Notice that R is a CM ring. We are concerned with the category $\mathfrak{C}(R)$ of CM modules over R. For this we only consider the case when $f \neq 0$, for otherwise, R will be a regular local ring on which all CM modules are free.

Let us begin by analyzing free resolutions of CM modules. For any CM module M over R, regarding it as an S-module, we have, by the Auslander-Buchsbaum formula, the following equality:

$$\text{proj.dim}_S(M) = \text{depth}(S) - \text{depth}_S(M) = 1.$$

Equivalently M has the following free resolution as an S-module:

(7.1.1) $$0 \longrightarrow S^{(n)} \xrightarrow{\varphi} S^{(n)} \longrightarrow M \longrightarrow 0.$$

(The middle and the left term have the same rank n, because M has rank 0 as an S-module.) Since M is an R-module, it follows that $fM = 0$. In particular, we see from (7.1.1) that $fS^{(n)} \subset \varphi(S^{(n)})$, thus for any x in $S^{(n)}$ there is a unique element $y \in S^{(n)}$ with $f \cdot x = \varphi(y)$. Putting $y = \psi(x)$, clearly ψ is a linear mapping from $S^{(n)}$ into itself and satisfies

(7.1.2) $$\varphi \cdot \psi = f \cdot 1_{S^{(n)}}.$$

Multiplying by φ from the right on both sides of (7.1.2), we have, since φ is a monomorphism,

(7.1.3) $$\psi \cdot \varphi = f \cdot 1_{S^{(n)}}.$$

We regard φ, ψ as square matrices on S by fixing a base of $S^{(n)}$.

(7.1) DEFINITION. A pair of square matrices (φ, ψ) with entries in S satisfying the conditions (7.1.2) and (7.1.3) is called a **matrix factorization** of f. A **morphism** between matrix factorizations (φ_1, ψ_1) and (φ_2, ψ_2) is a pair of matrices (α, β) with $\alpha \cdot \varphi_1 = \varphi_2 \cdot \beta$ and $\beta \cdot \psi_1 = \psi_2 \cdot \alpha$:

(7.1.4)
$$\begin{array}{ccccc} S^{(n_1)} & \xrightarrow{\psi_1} & S^{(n_1)} & \xrightarrow{\varphi_1} & S^{(n_1)} \\ \alpha \downarrow & & \beta \downarrow & & \alpha \downarrow \\ S^{(n_2)} & \xrightarrow{\psi_2} & S^{(n_2)} & \xrightarrow{\varphi_2} & S^{(n_2)} \end{array}$$

In this case, we write $(\alpha, \beta) : (\varphi_1, \psi_1) \to (\varphi_2, \psi_2)$. Note that the commutativity of the right square in (7.1.4) implies the commutativity of the left. In fact, multiplying $\alpha \cdot \varphi_1 = \varphi_2 \cdot \beta$ by ψ_1, ψ_2, we will have $f\psi_2 \cdot \alpha = \psi_2 \cdot \alpha \cdot \varphi_1 \cdot \psi_1 = \psi_2 \cdot \varphi_2 \cdot \beta \cdot \psi_1 = f\beta \cdot \psi_1$, hence $\psi_2 \cdot \alpha = \beta \cdot \psi_1$. For practical use, *the equality $\alpha \cdot \varphi_1 = \varphi_2 \cdot \beta$ is enough for (α, β) to be a morphism*. We denote by $\mathrm{MF}_S(f)$ the category of matrix factorizations of f and morphisms between them. It is trivial that $\mathrm{MF}_S(f)$ is an additive category, by defining the direct sum as follows:

$$(\varphi_1, \psi_1) \oplus (\varphi_2, \psi_2) = \left(\begin{pmatrix} \varphi_1 & 0 \\ 0 & \varphi_2 \end{pmatrix}, \begin{pmatrix} \psi_1 & 0 \\ 0 & \psi_2 \end{pmatrix} \right).$$

Naturally two matrix factorizations (φ_1, ψ_1) and (φ_2, ψ_2) are called **equivalent** if α, β are isomorphisms in (7.1.4). We often identify equivalent matrix factorizations. A matrix factorization (φ, ψ) of f is called **reduced** if the entries of φ, ψ are nonunits. This is equivalent to saying that any matrix factorization equivalent to (φ, ψ) have nonunit entries. Note that $(1, f)$ and $(f, 1)$ are matrix factorizations of f which are not reduced.

WARNING. Do not confuse (φ, ψ) with (ψ, φ). In general, they are not equivalent to each other; see (7.7).

If $(\varphi, \psi) \in \mathrm{MF}_S(f)$, then φ and ψ must each be a monomorphism as a mapping on $S^{(n)}$. In fact, if $\varphi(x) = 0$, then $fx = \psi(\varphi(x)) = 0$ hence $x = 0$, because f is a nonzero divisor on $S^{(n)}$.

Let (φ, ψ) be a matrix factorization of f. Denoting by $\overline{\varphi}, \overline{\psi}$ the matrices φ, ψ modulo (f), we have a chain complex of R-modules:

$$(7.2.1) \qquad \cdots \longrightarrow R^{(n)} \xrightarrow{\overline{\varphi}} R^{(n)} \xrightarrow{\overline{\psi}} R^{(n)} \xrightarrow{\overline{\varphi}} R^{(n)} \xrightarrow{\overline{\psi}} R^{(n)} \xrightarrow{\overline{\varphi}} R^{(n)} \xrightarrow{\overline{\psi}} R^{(n)} \longrightarrow \cdots.$$

This is actually a chain complex by (7.1.2) and (7.1.3). In fact,

(7.2.2)   the complex (7.2.1) is exact.

Indeed, if $\overline{z} \in R^{(n)}$ ($z \in S^{(n)}$) satisfies $\overline{\varphi}(\overline{z}) = 0$, then $\varphi(z) \in fS^{(n)} = \varphi \cdot \psi(S^{(n)})$, hence $z \in \psi(S^{(n)})$, for φ is a monomorphism. It thus follows that $\overline{z} \in \mathrm{Im}(\overline{\psi})$.

Thus one can obtain the R-free resolution of M by (7.1.1) together with (7.2.2):

$$(7.2.3) \qquad \cdots \longrightarrow R^{(n)} \xrightarrow{\overline{\varphi}} R^{(n)} \xrightarrow{\overline{\psi}} R^{(n)} \xrightarrow{\overline{\varphi}} R^{(n)} \xrightarrow{\overline{\psi}} R^{(n)} \xrightarrow{\overline{\varphi}} R^{(n)} \longrightarrow M \longrightarrow 0.$$

One observes here a curious phenomenon on the R-free resolutions of CM modules, which can be stated as follows:

(7.2) PROPOSITION. *A CM module over a hypersurface has a periodic free resolution with periodicity 2.*

We showed that if we are given a nontrivial CM module over a hypersurface $R = S/(f)$, then we have a matrix factorization (φ, ψ) of f and an R-free resolution as in (7.2.3). Conversely a matrix factorization (φ, ψ) will give a short exact sequence of S-modules as in (7.1.1) with M a CM module over R. In fact, because $fS^{(n)} \subset \varphi(S^{(n)})$, we have $fM = 0$, hence M is an R-module and M is equal to $\mathrm{Coker}(\varphi)$ which has an R-free resolution (7.2.3). By the periodicity of (7.2.3) M is the $2i$-th syzygy of M itself for any i, thus it must be a CM module by (1.16). We write $\mathrm{Coker}(\varphi, \psi)$ for M. Likewise, if $(\alpha, \beta) : (\varphi_1, \psi_1) \to (\varphi_2, \psi_2)$ is a morphism in $\mathrm{MF}_S(f)$, then it induces a chain map:

$$\begin{array}{ccccccccccc}
\cdots \xrightarrow{\overline{\psi_1}} & R^{(n_1)} & \xrightarrow{\overline{\varphi_1}} & R^{(n_1)} & \xrightarrow{\overline{\psi_1}} & R^{(n_1)} & \xrightarrow{\overline{\varphi_1}} & R^{(n_1)} & \longrightarrow & \mathrm{Coker}(\varphi_1, \psi_1) & \longrightarrow 0 \\
& \overline{\beta}\downarrow & & \overline{\alpha}\downarrow & & \overline{\beta}\downarrow & & \overline{\alpha}\downarrow & & & \\
\cdots \xrightarrow{\overline{\psi_2}} & R^{(n_2)} & \xrightarrow{\overline{\varphi_2}} & R^{(n_2)} & \xrightarrow{\overline{\psi_2}} & R^{(n_2)} & \xrightarrow{\overline{\varphi_2}} & R^{(n_2)} & \longrightarrow & \mathrm{Coker}(\varphi_2, \psi_2) & \longrightarrow 0
\end{array}$$

Hence there is a homomorphism of CM modules $\mathrm{Coker}(\varphi_1, \psi_1) \to \mathrm{Coker}(\varphi_2, \psi_2)$ which we will denote by $\mathrm{Coker}(\alpha, \beta)$. Thus we have defined an additive functor $\mathrm{Coker} : \mathrm{MF}_S(f) \to \mathfrak{C}(R)$. Clearly $\mathrm{Coker}(1, f) = 0$ and $\mathrm{Coker}(f, 1) = R$.

Before stating our result, we need some more preparation.

(7.3) DEFINITION. Let $\mathfrak{A}$ be a category whose Hom-sets are Abelian groups, and let $\mathfrak{P}$ be a set of objects in $\mathfrak{A}$. In general, we define $\mathfrak{A}/\mathfrak{P}$ to be the category whose objects are the same as $\mathfrak{A}$, and morphisms from A to B in $\mathfrak{A}/\mathfrak{P}$ are the elements of

$$\operatorname{Hom}_{\mathfrak{A}}(A,B)/\mathfrak{P}(A,B),$$

where $\mathfrak{P}(A, B)$ is a subgroup generated by all morphisms from A to B which pass through direct sums of objects in $\mathfrak{P}$. Notice that any objects in $\mathfrak{P}$ are the zero object in $\mathfrak{A}/\mathfrak{P}$ and that $\mathfrak{A}/\mathfrak{P}$ is the largest quotient of $\mathfrak{A}$ with this property.

Using this notation, we define the categories:

(7.3.1) $\quad\underline{\mathrm{MF}}_S(f) \ = \mathrm{MF}_S(f)/\{(1,f)\},$

(7.3.2) $\quad\underline{\mathrm{RMF}}_S(f) = \mathrm{MF}_S(f)/\{(1,f),(f,1)\},$

(7.3.3) $\quad\underline{\mathfrak{C}}(R) \quad\ = \mathfrak{C}(R)/\{R\}.$

Note that $\underline{\mathfrak{C}}(R)$ is the category having $\underline{\operatorname{Hom}}_R(A, B)$ as the set of morphisms from A to B, cf. (3.7).

We can now state the theorem.

(7.4) THEOREM. (Eisenbud's matrix factorization theorem [25]) *If $R = S/(f)$ is a hypersurface, then* Coker *induces an equivalence of the categories:*

$$\underline{\mathrm{MF}}_S(f) \simeq \underline{\mathfrak{C}}(R).$$

Moreover this induces an equivalence:

$$\underline{\mathrm{RMF}}_S(f) \simeq \underline{\mathfrak{C}}(R).$$

PROOF: Since $\mathrm{Coker}(1, f) = 0$, Coker certainly induces the functor $\underline{\mathrm{MF}}_S(f) \to \underline{\mathfrak{C}}(R)$, which we also denote by Coker. Define the functor $\Upsilon : \underline{\mathfrak{C}}(R) \to \underline{\mathrm{MF}}_S(f)$ as follows: For a nontrivial CM module M we may have a free resolution of M as in (7.1.1), and then obtain matrices (φ, ψ) as in (7.1.2) and (7.1.3). We set $\Upsilon(M) = (\varphi, \psi)$. Note that this is determined uniquely as an object in $\underline{\mathrm{MF}}_S(f)$. In fact, if we choose φ so that the resolution (7.1.1) is minimal and if (φ_1, ψ_1) is another matrix factorization obtained from M, then there are invertible matrices α and β such that the following diagram is commutative:

$$\begin{array}{ccccccccc}
0 & \longrightarrow & S^{(n_1)} & \xrightarrow{\begin{pmatrix}\varphi & 0\\ 0 & 1\end{pmatrix}} & S^{(n_1)} & \longrightarrow & M & \longrightarrow & 0 \\
& & \beta\downarrow & & \alpha\downarrow & & \parallel & & \\
0 & \longrightarrow & S^{(n_1)} & \xrightarrow{\varphi_1} & S^{(n_1)} & \longrightarrow & M & \longrightarrow & 0
\end{array}$$

Hence (α, β) is a morphism from $\left(\begin{pmatrix} \varphi & 0 \\ 0 & 1 \end{pmatrix}, \begin{pmatrix} \psi & 0 \\ 0 & f \cdot 1 \end{pmatrix}\right)$ to (φ_1, ψ_1) which gives an equivalence. Since we may neglect $(1, f)$ in $\underline{\mathrm{MF}}_S(f)$, $\Upsilon(M) \in \underline{\mathrm{MF}}_S(f)$ is uniquely determined.

Next we consider morphisms. Given a morphism $g : M_1 \to M_2$ in $\mathfrak{C}(R)$, there is a commutative diagram:

(7.4.1)
$$\begin{array}{ccccccccc} 0 & \longrightarrow & S^{(n_1)} & \xrightarrow{\varphi_1} & S^{(n_1)} & \longrightarrow & M_1 & \longrightarrow & 0 \\ & & \beta \downarrow & & \alpha \downarrow & & g \downarrow & & \\ 0 & \longrightarrow & S^{(n_2)} & \xrightarrow{\varphi_2} & S^{(n_2)} & \longrightarrow & M_2 & \longrightarrow & 0. \end{array}$$

Hence (α, β) gives a morphism $\Upsilon(M_1) \to \Upsilon(M_2)$ which we denote by $\Upsilon(g)$. If (α', β') is another pair of matrices which makes (7.4.1) commutative, then there is $\mu : S^{(n_1)} \to S^{(n_2)}$ such that $\alpha - \alpha' = \varphi_2 \cdot \mu$ and $\beta - \beta' = \mu \cdot \varphi_1$, therefore the morphism $(\alpha, \beta) - (\alpha', \beta')$ is a composition of $(\mu, \mu \cdot \varphi_1) : (\varphi_1, \psi_1) \to (1_{n_2}, f \cdot 1_{n_2})$ with $(\varphi_2, 1) : (1_{n_2}, f \cdot 1_{n_2}) \to (\varphi_2, \psi_2)$. Hence $(\alpha, \beta) = (\alpha', \beta')$ as a morphism in $\underline{\mathrm{MF}}_S(f)$. Consequently $\Upsilon(g)$ is uniquely determined. We thus have defined the functor $\Upsilon : \mathfrak{C}(R) \to \underline{\mathrm{MF}}_S(f)$. Then by the definitions of Coker and Υ it is fairly easy to see that $\Upsilon \cdot \mathrm{Coker} = 1$ and $\mathrm{Coker} \cdot \Upsilon = 1$. We therefore obtain an equivalence $\underline{\mathrm{MF}}_S(f) \simeq \mathfrak{C}(R)$. The second equivalence in the theorem is obvious from the first, since $\mathrm{Coker}(f, 1) = R$. ∎

(7.5) REMARK. Let (φ, ψ) be an object in $\mathrm{MF}_S(f)$. Notice first that

(7.5.1) if (φ, ψ) is a reduced matrix factorization, then $\mathrm{Coker}(\varphi, \psi)$ has no free summand.

In fact if M has a summand R, then in (7.1.1) φ may be taken as $\varphi' \oplus f$ for some φ', hence (φ, ψ) is equivalent to $(\varphi', \psi') \oplus (f, 1)$ for some ψ', which is a contradiction.

If the matrix φ has a unit entry, then it is easy to see that (φ, ψ) has $(1, f)$ as a summand. Likewise, if ψ contains a unit, then (φ, ψ) has $(f, 1)$ as a summand. As a result, an arbitrary matrix factorization (φ, ψ) can be written as

(7.5.2) $$(\varphi, \psi) = (\varphi_0, \psi_0) \oplus (f, 1)^{(p)} \oplus (1, f)^{(q)},$$

with (φ_0, ψ_0) reduced and with p, q nonnegative integers. We claim that *this decomposition is unique up to equivalence.* To show this, let $(\varphi, \psi) = (\varphi'_0, \psi'_0) \oplus (f, 1)^{(p')} \oplus (1, f)^{(q')}$ be another one. Putting $M = \mathrm{Coker}(\varphi, \psi)$, $M_0 = \mathrm{Coker}(\varphi_0, \psi_0)$ and $M'_0 = \mathrm{Coker}(\varphi'_0, \psi'_0)$, we see from (7.4) that $M \simeq M_0 \oplus R^{(p)} \simeq M'_0 \oplus R^{(p')}$. Since, by (7.5.1), M_0 and M'_0 have no free summands, we have $p = p'$ and $M_0 \simeq M'_0$ by the uniqueness of direct decompositions in $\mathfrak{C}(R)$. From $M_0 \simeq M'_0$, one sees that (φ_0, ψ_0) is equivalent to (φ'_0, ψ'_0); see the construction of Υ in the proof of (7.4). Finally, comparing the sizes of matrices, one obtains $q = q'$ and the uniqueness of (7.5.2) follows.

By this remark we see that under the equivalence in (7.4), the reduced matrix factorizations correspond precisely to CM modules with no free summand. Thus we have:

(7.6) COROLLARY. *The functor* Coker *yields a bijective correspondence between the set of equivalence classes of reduced matrix factorizations of f and the set of isomorphism classes of CM modules over R which have no free summands.*

Obviously an indecomposable matrix factorization corresponds to an indecomposable CM module.

(7.7) PROPOSITION. *Let M be an indecomposable nonfree CM module over $R = S/(f)$ given by $M = \operatorname{Coker}(\varphi, \psi)$ with (φ, ψ) a reduced matrix factorization of f. Then $\operatorname{syz}_R^1 M$ is also indecomposable and nonfree, and $\operatorname{syz}_R^1 M \simeq \operatorname{Coker}(\psi, \varphi)$.*

PROOF: Recall, (1.15), that $\operatorname{syz}_R^1 M$ denotes the reduced first syzygy of M. It is obvious that (ψ, φ) is also in $\mathrm{MF}_S(f)$. Furthermore it is indecomposable, for otherwise, (ψ, φ) would be equivalent to a nontrivial sum $(\psi_1, \varphi_1) \oplus (\psi_2, \varphi_2)$, and then (φ, ψ) would be equivalent to $(\varphi_1, \psi_1) \oplus (\varphi_2, \psi_2)$, a contradiction. Note that (ψ, φ) is reduced and hence $\operatorname{Coker}(\psi, \varphi)$ is nonfree. From the exact sequence (7.2.1) and from the definition of Coker it follows that there is an exact sequence:

$$0 \longrightarrow \operatorname{Coker}(\psi, \varphi) \longrightarrow R^{(n)} \longrightarrow \operatorname{Coker}(\varphi, \psi) \longrightarrow 0.$$

Since $\operatorname{Coker}(\psi, \varphi)$ is indecomposable nonfree, we have $\operatorname{syz}_R^1 M \simeq \operatorname{Coker}(\psi, \varphi)$ as desired. ∎

(7.8) REMARK. Let M be the same as in (7.7) and let N be another CM module over R given by $N = \operatorname{Coker}(\varphi', \psi')$ for a matrix factorization (φ', ψ'). If $h : N \to \operatorname{syz}_R^1 M$ is an R-homomorphism, then by (7.4) there is a morphism $(\alpha, \beta) : (\varphi', \psi') \to (\psi, \varphi)$ in $\mathrm{MF}_S(f)$ such that $\operatorname{Coker}(\alpha, \beta) = h$. By the definition of morphisms, it is easy to see that $(\begin{pmatrix} \varphi & \beta \\ 0 & \varphi' \end{pmatrix}, \begin{pmatrix} \psi & -\alpha \\ 0 & \psi' \end{pmatrix})$ is also a matrix factorization of f. Letting $L = \operatorname{Coker}(\begin{pmatrix} \varphi & \beta \\ 0 & \varphi' \end{pmatrix}, \begin{pmatrix} \psi & -\alpha \\ 0 & \psi' \end{pmatrix})$, we have an exact sequence:

$$0 \longrightarrow M \longrightarrow L \longrightarrow N \longrightarrow 0,$$

whose class in $\operatorname{Ext}_R^1(N, M)$ is the image of h under the natural mapping $\rho : \operatorname{Hom}_R(N, \operatorname{syz}_R^1 M) \to \operatorname{Ext}_R^1(N, M)$. Since ρ is surjective, any extension of M by N is obtained in this way.

Chapter 8. Simple singularities

In this chapter we shall define simple singularities in the sense of commutative algebra and show that any Gorenstein ring of finite representation type has a simple singularity; see Theorem (8.10) and Corollary (8.16). The converse is also true. However we postpone its proof to the next chapter, for it requires a more complicated argument. Almost all arguments below are taken from the papers Buchweitz-Greuel-Schreyer [19] and Herzog [35].

Throughout this chapter $(R, \mathfrak{m}, k)$ is a Henselian CM local ring. We always assume that R is a hypersurface defined by f in a regular local ring S:

$$R = S/(f)$$

The maximal ideal of S is always denoted by $\mathfrak{n}$.

(8.1) DEFINITION. For a hypersurface $R = S/(f)$ consider the following set of ideals in S:

$$c(f) = \{ I \mid I \text{ is a proper ideal of } S \text{ with } f \in I^2 \}.$$

Call R a **local ring of simple singularity** (or a **simple hypersurface singularity**) if the set $c(f)$ is finite.

We first show how this definition regulates the form of f.

(8.2) LEMMA. *Let $R = S/(f)$ be a local ring of simple singularity and assume that k is algebraically closed. Then the following hold:*
(8.2.1)   *If $\dim(R) = 1$, then $e(R) \leq 3$.*
(8.2.2)   *If $\dim(R) \geq 2$, then $e(R) \leq 2$,*
where $e(R)$ denotes the multiplicity of R, that is, $e(R)$ is the maximal integer e with $f \in \mathfrak{n}^e$.

PROOF: Let $\pi : \mathfrak{n} \to \mathfrak{n}/\mathfrak{n}^2$ be the natural projection and let $\{x_0, x_1, \ldots, x_d\}$ be a set of elements in $\mathfrak{n}$ with $\{\pi(x_0), \pi(x_1), \ldots, \pi(x_d)\}$ a k-base of $\mathfrak{n}/\mathfrak{n}^2$, where $d = \dim(R)$.

First suppose $e(R) \geq 4$. Then letting $J_\lambda = \pi^{-1}(\lambda)$ for any subspace λ of $\mathfrak{n}/\mathfrak{n}^2$, we deduce that $J_\lambda \neq J_{\lambda'}$ when $\lambda \neq \lambda'$ and that $f \in J_\lambda^2$ for any λ, since $f \in \mathfrak{n}^4$. If $d \geq 1$, then there are infinitely many subspaces in $\mathfrak{n}/\mathfrak{n}^2$, hence R is not a simple singularity. This proves (8.2.1).

Next suppose $e(R) \geq 3$ and $d \geq 2$. Then f can be written as an infinite sum:

$$f = \sum_{i \geq 3} f_i,$$

where each f_i is a homogeneous polynomial of degree i in $\{x_0, x_1, ..., x_d\}$. Let V be a hypersurface in $\mathbb{P}_k^d$ defined by the equation $f_3 = 0$. Note that V has dimension at least 1, since $d \geq 2$. Hence it is an infinite set. For any point $\lambda = (a_0 : a_1 : \ldots : a_d)$ in V, we define an ideal I_λ as $\{a_i x_j - a_j x_i|\ 0 \leq i, j \leq d\}S + \mathfrak{n}^2$. It is obvious that $I_\lambda \neq I_{\lambda'}$ if $\lambda \neq \lambda'$, because the homogeneous equations in $I_\lambda/\mathfrak{n}^2$ exactly define the point λ in $\mathbb{P}_k^d$. Thus the lemma will follow if we prove $f \in I_\lambda^2$ ($\lambda \in V$). To see this we may assume $\lambda = (1 : 0 : \ldots : 0)$ after a change of basis. Then $I_\lambda = (x_0^2, x_1, \ldots, x_d)S$, so we have $I_\lambda^2 = \{x_i x_j, x_0^2 x_i, x_0^4|\ 1 \leq i, j \leq d\}S$. It hence follows that $f_3 \equiv \alpha x_0^3 \pmod{I_\lambda^2}$ with $\alpha \in k$. Since λ is a point in V, we see that $\alpha = f_3(1, 0, \ldots, 0) = 0$, which shows $f_3 \in I_\lambda^2$. Therefore $f \in I_\lambda^2 + \mathfrak{n}^4 = I_\lambda^2$. ∎

(8.3) LEMMA. *Let $R = S/(f)$ be a simple singularity of positive dimension. Then R is reduced.*

PROOF: Suppose R contains a nilpotent element. Then we have a decomposition $f = g \cdot h^2$ for some $g, h \in S$. It is then obvious that $f \in I^2$ for any ideal I of S containing h. Since $S/(h)$ has dimension at least one, there are infinitely many such I. This shows the lemma. ∎

(8.4) LEMMA. *Let $R = S/(f)$ be a simple singularity of dimension 1. Assume that S contains an infinite field F. Then for any two elements x, y in $\mathfrak{n}$ we have:*

(8.4.1) $$f \notin (x^3, x^2 y^2, x y^4, y^6).$$

PROOF: Assume the contrary. Then $f \in (x^3, x^2 y^2, xy^4, y^6)$ for some $x, y \in \mathfrak{n}$. Considering the set of ideals in S:

$$I_\lambda = (x + \lambda y^2, y^3) \quad (\lambda \in F),$$

we can show that $I_\lambda \neq I_{\lambda'}$ if $\lambda \neq \lambda'$. In fact, if $I = I_\lambda = I_{\lambda'}$ for some $\lambda \neq \lambda'$, then $(\lambda - \lambda') y^2 \in I$, hence I contains y^2 and x. Thus $I = (x, y^2)$. Considering the image of (x, y^2) (resp. $(x + \lambda y^2, y^3)$) in $S/(x + \lambda y^2)S$, it is generated by y^2 (resp. y^3). Therefore $y^2 - ay^3 = y^2(1 - ay) \in (x + \lambda y^2)S$ for some $a \in S$, thus $y^2 \in (x + \lambda y^2)S$ and hence

$x \in (x + \lambda y^2)S$. Then by the assumption, $f \in (x^3, x^2y^2, xy^4, y^6) = I^3 \subseteq (x + \lambda y^2)^3 S$, contradicting (8.3). We have thus shown that there are infinitely many I_λ. If we see that $f \in I_\lambda^2$, then the proof will be finished. This is, however, trivial, since $(y^6, xy^4, x^2y^2, x^3) \subseteq I_\lambda^2$. ∎

(8.5) PROPOSITION. *Let $S = k\{x, y\}$ where k is an algebraically closed field of characteristic 0. If $R = S/(f)$ is a simple singularity, then after a change of variables f is equal to one of the following polynomials:*

(A_n)	$x^2 + y^{n+1}$	$(n \geq 1)$,
(D_n)	$x^2 y + y^{n-1}$	$(n \geq 4)$,
(E_6)	$x^3 + y^4$,	
(E_7)	$x^3 + xy^3$,	
(E_8)	$x^3 + y^5$.	

Before the proof of the proposition we need a lemma.

(8.6) LEMMA. *Let S be a convergent power series ring $k\{x_0, x_1, \ldots, x_d\}$ over an algebraically closed field k of characteristic 0. Then for any unit element u in S and for any positive integer n, there is a power series v in S such that $v^n = u$. In particular, there is a k-algebra automorphism of S which sends x_0^n to ux_0^n and sends each x_i to itself for $i \geq 1$.*

PROOF: Considering the algebraic equation $X^n - u \equiv 0 \mod (x_0, x_1, \ldots, x_d)S$, we see that it has n distinct solutions in $k = S/(x_0, x_1, \ldots, x_d)S$. Since S is a Henselian ring, it follows that the equation $X^n - u = 0$ has a solution in S, showing the existence of v. The last statement of the lemma follows from the inverse function theorem. In fact, defining the k-algebra map $\varphi : S \to S$ by $\varphi(x_0) = vx_0$ and $\varphi(x_i) = x_i$ $(i \geq 1)$, we can see that φ is an automorphism. ∎

Before the proof of Proposition (8.5) we note the following fact:
For the ring R as in (8.5), a general element y in $\mathfrak{m} - \mathfrak{m}^2$ satisfies $y\mathfrak{m}^n = \mathfrak{m}^{n+1}$ for some integer n. Such an element is called a **minimal reduction** of $\mathfrak{m}$ and it satisfies;

(8.6.1) R is module-finite over a subalgebra $T = k\{y\}$ and there is an isomorphism of T-algebras $R \simeq T[X]/(g(X))$ with $\deg(g(X)) = e(R)$. (Matsumura [48].)

PROOF OF (8.5): We divide the proof into several cases.
(i) *The case $e(R) = 2$:* Take a minimal reduction y of the maximal ideal $\mathfrak{m}$ of R. Then we may describe as $R = T[x]/(x^2 + a)$ for some $a \in T$. Thus we may assume that

$f = x^2 + a$. Write a as uy^{n+1} with u a unit in T. Then by (8.6), after applying a suitable automorphism on S, we can take $f = x^2 + y^{n+1}$. This is the case (A_n).

(ii) *The case $e(R) = 3$ with $\{f = 0\}$ has two or three different tangent directions*: In this case f is decomposed as the product $g \cdot h$ $(g, h \in S)$ where $\{g = 0\}$ and $\{h = 0\}$ have distinct tangents. Since $e(R) = e(S/(g)) + e(S/(h))$, we may assume that $e(S/(g)) = 2$ and $e(S/(h)) = 1$. We can take $y \in R$ whose images in $S/(g)$ and in $S/(h)$ are minimal reductions of their maximal ideals. As in the first case we may have $S/(g) \simeq T[X]/(X^2 + y^n)$ where $T = k\{y\}$. Then $R \simeq T[X]/((X-t)(X^2+y^n))$ $(t \in T)$, therefore we may assume that $f = (x-t)(x^2+y^n)$ $(n \geq 2)$. Suppose $n \geq 3$. Since $\{f = 0\}$ has different tangent directions, we see that $t = uy$ with u a unit in T. Replacing y by $x - uy$ we can change f into the form $y(ax^2 + bxy^{n-1} + cy^n)$ $(a, b, c$ units in $S)$. Then putting $\xi = a^{\frac{1}{2}}x + \frac{1}{2}a^{-\frac{1}{2}}by^{n-1}$ and $\eta = y(c - \frac{1}{4}a^{-1}b^2y^{n-2})^{\frac{1}{n}}$, we have $f = \eta(\xi^2 + \eta^n)$ up to a unit; this is the case (D_{n+2}) $(n \geq 3)$. We leave the reader the proof for the case $n = 2$. In this case f could be chosen as $y(x^2 + y^2)$ that is the equation of (D_4).

(iii) *The case $e(R) = 3$ with $f = 0$ has a unique tangent direction*: We further divide this into two cases.

(iii − 1) *The case when f is reducible*: Taking y as a minimal reduction and putting $T = k\{y\}$, we may assume that $f = x(x^2 + ax + b)$ $(a, b \in T)$. Since x^3 should be the initial form of f, we may write $f = x(x^2 + cxy^2 + dy^3)$ $(c, d \in T)$, where d must be a unit in T because of (8.4). Then using (8.6), f can be put into the form $x(x^2 + exy^2 + y^3)$ $(e \in T)$. We replace y by $y - \frac{1}{3}ex$ to get

$$f = x(x^2 + y^3 + sx^2y + tx^3),$$

for some $s, t \in S$. Changing $x(1 + sy + tx)^{\frac{1}{2}}$ to x, we finally get $f = x(x^2 + y^3)$, the equation of (E_7), up to a unit.

(iii−2) *The case when f is irreducible*: We may write $R = k\{y\}[X]/(X^3 + aX + b)$ $(a, b \in T = k\{y\})$, hence we may assume that $f = x^3 + ax + b$. Since R is an integral domain, we can see that $b \neq 0$. If $a = 0$ then, by (8.6), f can be put into the form $x^3 + y^m$, where $m = 4$ or 5 by (8.4). These are the cases (E_6) and (E_8). So we assume that $a \neq 0$. Then we may write

(*) $$f = x^3 + uxy^n + y^m,$$

where $u \in T$ is a unit element. Since x^3 must be the initial form of f, we see that $n \geq 3$ and $m \geq 4$. We claim here that either $n \geq 4$ or $m = 4$. Suppose not. Then we could find a solution $\xi \in T$ to the equation; $\xi^3 + u\xi^2 + y^{(2m-9)} = 0$. In fact, modulo y, the equation has a simple root $-u$, and it can be lifted to a solution in T, since T is a Henselian ring. Furthermore ξ is a unit in T as well as u. Then it would follow that $f(\xi^{-1}y^{m-3}, y) = 0$ in (*). (Use $n = 3$ here.) Hence f would be divisible by $x - \xi^{-1}y^{m-3}$, which contradicts that f is irreducible. Hence we have shown that either $n \geq 4$ or $m = 4$.

If $m = 4$ in (*), then replacing y by $y - \frac{1}{4}uxy^{n-3}$ and using (8.6) we have $f = x^3 + wy^2x^2 + y^4$ for some $w \in S$, and a further change of variables $x \mapsto x + \frac{1}{3}wy^2$ takes f into the form $x^3 + ey^4$ (e is a unit). Therefore by (8.6) we may have $f = x^3 + y^4$, which is the case (E_6).

If $n \geq 4$ and $m \neq 4$, then by (8.4) we see that $m = 5$. Replacing y by $y + \frac{1}{5}ux$ and using (8.6) we have $f = x^3 + wy^3x^2 + y^5$ for some $w \in S$, and then by a further change of variables $x \mapsto x + \frac{1}{3}wy^3$, f can be put into the form $x^3 + ey^5$ (e is a unit). Finally, by (8.6), we have $f = x^3 + y^5$, which is the case (E_8). This completes the proof. ∎

(8.7) *Exercise.* Prove the converse of the proposition. Namely, if f is one of the polynomials in (8.5), then the ring $R = S/(f)$ is a simple singularity. In this case describe the set $c(f)$ for each polynomial.

(8.8) THEOREM. *Let $S = k\{x, y, z_2, z_3, \ldots, z_d\}$ and assume that k is an algebraically closed field of characteristic 0. If $R = S/(f)$ is a simple singularity, then after a suitable change of variables, f is equal to one of the following polynomials:*

(A_n)	$x^2 + y^{n+1} + z_2^2 + z_3^2 + \ldots + z_d^2$	$(n \geq 1)$,
(D_n)	$x^2y + y^{n-1} + z_2^2 + z_3^2 + \ldots + z_d^2$	$(n \geq 4)$,
(E_6)	$x^3 + y^4 + z_2^2 + z_3^2 + \ldots + z_d^2$,	
(E_7)	$x^3 + xy^3 + z_2^2 + z_3^2 + \ldots + z_d^2$,	
(E_8)	$x^3 + y^5 + z_2^2 + z_3^2 + \ldots + z_d^2$.	

PROOF: Note that d is the dimension of the ring R. If $d = 1$, then the theorem is nothing but the previous proposition. So we may assume $d \geq 2$. In this case we know by (8.2) that $e(R) = 2$. In other words $f \in \mathfrak{n}^2 - \mathfrak{n}^3$. Then by the Weierstrass preparation theorem, after changing variables, one can write

$$f(x, y, z_2, z_3, \ldots, z_d) = g(x, y, z_2, z_3, \ldots, z_{d-1}) + z_d^2,$$

where g is in $S' = k\{x, y, z_2, z_3, \ldots, z_{d-1}\}$. Here it can be seen that $S'/(g)$ is also a simple singularity. In fact it is an easy exercise to show that the following mapping is injective:

$$\begin{array}{ccc} c(g) & \longrightarrow & c(f) \\ I & \longmapsto & (I, z_d). \end{array}$$

Hence the theorem is proved by induction on d. ∎

(8.9) *Exercise.* Prove the converse of the theorem. (Hint: If f is one of the polynomials in (8.8) and if $f \in I^2$, then $2z_i = \partial f/\partial z_i \in I$. Thus the exercise follows from (8.7).)

It is our main purpose here to prove the following theorem.

(8.10) THEOREM. (Buchweitz-Greuel-Schreyer [19]) *If a hypersurface $R = S/(f)$ is of finite representation type, then R is a simple singularity.*

To prove this we need some more preliminaries.

(8.11) DEFINITION. Let $\varphi : S^{(n)} \to S^{(n)}$ be a homomorphism of free S-modules. Then define $\Phi : S^{(n)} \otimes_S (S^{(n)})^* \to S$ to be $\Phi(f \otimes g) = g(\varphi(f))$ ($f \in S^{(n)}, g \in (S^{(n)})^*$) and denote the image of Φ by $I(\varphi)$. Note that if we write φ as a square matrix of size n by fixing a base of $S^{(n)}$, then $I(\varphi)$ is an ideal of S generated by all the entries in the matrix.

Let M be a CM module over R without free summands and let (φ, ψ) be a reduced matrix factorization of f corresponding to M, i.e. $M = \mathrm{Coker}(\varphi, \psi)$ with the notation in (7.4). Then define an ideal $I(M)$ of S as

$$I(M) = I(\varphi) + I(\psi).$$

Note that $I(M)$ is independent of the choice of (φ, ψ). In fact, if (φ', ψ') is another reduced matrix factorization of f with $M = \mathrm{Coker}(\varphi', \psi')$, then it must be equivalent to (φ, ψ) by (7.6). Then by the definition of equivalence, we have $I(\varphi) = I(\varphi')$ and $I(\psi) = I(\psi')$. Also note that $I(M \oplus N) = I(M) + I(N)$ for any CM modules M, N. It thus follows that if $\{M_\lambda | \lambda \in \Lambda\}$ is a complete set of indecomposable CM modules over R, then the set $\{I(M)|\ M$ is a CM module over $R\}$ equals $\{\sum_{\lambda \in \Delta} I(M_\lambda)|\ \Delta$ is a finite subset of $\Lambda\}$. In particular, if R has only a finite number of isomorphism classes of indecomposable CM modules, then this set is finite.

We evidently have the following lemma.

(8.12) LEMMA. *Let M be a CM module over a hypersurface $R = S/(f)$. Then $f \in I(M)^2$, that is, $I(M) \in c(f)$.*

PROOF: Let (φ, ψ) be a matrix factorization of f. Since $\varphi \cdot \psi = f \cdot 1_{S^{(n)}}$, we see that $f \in I(\varphi)I(\psi) \subset I(M)^2$. ∎

We can prove the following:

(8.13) LEMMA. *For a hypersurface $R = S/(f)$, consider I as a mapping from the set of classes of CM modules over R without free summands into $c(f)$. Then the mapping is surjective.*

Theorem (8.10) will be a straightforward consequence of this lemma. In fact if R has only a finite number of indecomposable CM modules, then the image of I is finite, hence $c(f)$ is a finite set by the lemma and R is a simple singularity.

We introduce some notation for the proof of Lemma (8.13).

Let $R = S/(f)$ be a hypersurface as above, where f belongs to $\mathfrak{n}^2$. Then f can be written as

(8.14.1) $$f = \sum_{i=1}^{r} x_i y_i \quad (x_i, y_i \in \mathfrak{n}).$$

Then we define linear maps on an exterior algebra $\bigwedge S^{(r)}$.

(8.14.2)   When $\{e_1, e_2, \ldots, e_r\}$ is a basis of $S^{(r)}$,

$$\delta_-(e_{i_1} \wedge e_{i_2} \wedge \ldots \wedge e_{i_t}) = \sum_{j=1}^{t} (-1)^{j-1} x_{i_j}(e_{i_1} \wedge \ldots \wedge \hat{e}_{i_j} \wedge \ldots \wedge e_{i_t}),$$

$$\delta_+(w) = (\sum_{j=1}^{r} y_j e_j) \wedge w.$$

There is thus no difficulty in showing that δ_+ and δ_- are differential maps on $\bigwedge S^{(r)}$ of degree respectively $+1$ and -1. ($\delta_+^2 = \delta_-^2 = 0$ and $\delta_\pm$ sends an exterior product of degree i to that of degree $i \pm 1$.) Setting $\delta = \delta_+ + \delta_-$, we see that

(8.14.3) $$\delta^2 = f \cdot 1_{(\bigwedge S^{(r)})}$$

Actually, since $\delta_+^2 = \delta_-^2 = 0$, we have $\delta^2 = \delta_+ \delta_- + \delta_- \delta_+$ and an easy computation shows $(\delta_+ \delta_- + \delta_- \delta_+)(e_{i_1} \wedge \ldots \wedge e_{i_t}) = f \cdot e_{i_1} \wedge \ldots \wedge e_{i_t}$. (Check this.)

By (8.14.2) and (8.14.3) we have shown:

(8.14) LEMMA. *With the above notation, (δ, δ) gives a matrix factorization of f and $I(\delta) = (x_1, x_2, \ldots, x_r, y_1, y_2, \ldots, y_r)S$.*

Now we prove Lemma (8.13).

Let I be an element of $c(f)$. We want to construct a CM module M over R with $I(M) = I$. Let I be generated by $\{x_1, x_2, \ldots, x_r\}$. Then f can be written as $f = \sum_{i=1}^{r} x_i y_i$ ($y_i \in I$), for $f \in I^2$. Constructing δ as in (8.14.2), we have a reduced matrix factorization (δ, δ) of f with the property $I(\delta) = I$. It is sufficient to take the CM module $\operatorname{Coker}(\delta, \delta)$ as M. This finishes the proof of (8.13) and hence (8.10). ∎

In Theorem (8.10) the assumption that R is a hypersurface is superfluous. Actually it is enough to assume R is a Gorenstein ring.

(8.15) THEOREM. (Herzog [35]) *Let R be a ring of the form $S/\mathfrak{a}$ where S is a regular local ring with maximal ideal $\mathfrak{n}$ and $\mathfrak{a}$ is an ideal of S with $\mathfrak{a} \subset \mathfrak{n}^2$. Assume that R is a Gorenstein ring of finite representation type. Then $\mathfrak{a}$ is a principal ideal, hence R is a hypersurface.*

By this theorem, (8.10) can be strengthened to:

(8.16) COROLLARY. *Assume that R is a Gorenstein ring and is a homomorphic image of a regular local ring. If R is of finite representation type, then R is a simple singularity.*

In order to prove (8.15) we need two lemmas.

(8.17) LEMMA. *Let R be a Gorenstein local ring and let*

$$0 \longrightarrow N \xrightarrow{q} F \xrightarrow{p} M \longrightarrow 0$$

be an exact sequence of CM R-modules, where F is a free R-module and $p \otimes R/\mathfrak{m}$ is an isomorphism. Then if M is indecomposable, so is N.

PROOF: Suppose N is decomposable, so that $N = N_1 \oplus N_2$ ($N_1 \neq 0, N_2 \neq 0$) and let q be (q_1, q_2) along this decomposition. Take the dual of the sequence by the canonical module to obtain an exact sequence:

$$0 \longrightarrow M' \longrightarrow F' \xrightarrow{\begin{pmatrix} q'_1 \\ q'_2 \end{pmatrix}} N'_1 \oplus N'_2 \longrightarrow 0.$$

Note that, since R is Gorenstein, F' is a free module, hence it is a free cover of $N'_1 \oplus N'_2$. First suppose that neither N_1 nor N_2 are free. In this case M' contains $\mathrm{syz}^1 N'_1 \oplus \mathrm{syz}^1 N'_2$ as a direct summand. Since $M'' \simeq M$, this contradicts that M is indecomposable. Secondly consider the case when N_1 is free. Then q'_1 must be a split epimorphism and hence q_1 is a split monomorphism. In particular, $q_1 \otimes R/\mathfrak{m}$ is a monomorphism. Since $p \cdot q_1 = 0$, this contradicts that $p \otimes R/\mathfrak{m}$ is an isomorphism. These contradictions show that N is indecomposable. ∎

Next we remark that the converse of (7.2) is true for Gorenstein rings.

(8.18) LEMMA. *Let R be $S/\mathfrak{a}$ as in (8.15). (S is a regular local ring and $\mathfrak{a} \subset \mathfrak{n}^2$.) If a free resolution of every CM module M over R is periodic except for a finite part, then $\mathfrak{a}$ is a principal ideal.*

Here we say that a free resolution

$$\cdots \longrightarrow G_{n+1} \xrightarrow{\varphi_n} G_n \longrightarrow \cdots \longrightarrow G_2 \xrightarrow{\varphi_1} G_1 \xrightarrow{\varphi_0} G_0 \longrightarrow M \longrightarrow 0$$

is **periodic except for a finite part** if there are positive integers n and h with $\varphi_{\mu+h} = \varphi_\mu$ for any $\mu \geq n$.

PROOF: Recall first the following: Suppose we are given a minimal free resolution of an R-module M:

$$\cdots \longrightarrow F_n \longrightarrow F_{n-1} \longrightarrow \cdots \longrightarrow F_1 \longrightarrow F_0 \longrightarrow M \longrightarrow 0.$$

We denote the rank of each F_n by $\beta_n(M)$ and call it the n-th **Betti number** of M. Tate [63] showed that $\mathfrak{a}$ is a principal ideal if and only if there is an upper bound for the set $\{\beta_n(k)|\ 1 \leq n < \infty\}$.

In the following we want to show that under the assumption of the lemma, there is a bound for the Betti numbers of k. To do this, consider the free resolution of k to get an exact sequence

$$0 \longrightarrow M \longrightarrow F_{d-1} \longrightarrow \cdots \longrightarrow F_2 \longrightarrow F_1 \longrightarrow F_0 = R \longrightarrow k \longrightarrow 0,$$

where d denotes the dimension of R. Note from (1.4) that M is a CM module over R. Decompose M into indecomposable modules as $M = \sum_j M_j$. Since the free resolutions of the M_j are periodic except for a finite part, there are bounds for the Betti numbers for the M_j. It, thus, follows that the set of Betti numbers of k also has a bound. ∎

Now we proceed to the proof of Theorem (8.15).

Let R be the same as in (8.15). It is, from (8.18), sufficient to show that any CM module M over R has a free resolution with periodicity except for a finite part. Clearly we may assume M is an indecomposable CM module. Consider the minimal free cover of M to have an exact sequence:

$$0 \longrightarrow N \longrightarrow F_0 \xrightarrow{p} M \longrightarrow 0.$$

Note that $p \otimes R/\mathfrak{m}$ is an isomorphism. We then know from (8.17) that N is also an indecomposable CM module. Write $\Phi(M)$ instead of N. Then Φ gives a mapping from the set of classes of indecomposable CM modules over R into itself. Since this is a finite set, it can easily be seen that there are positive integers n and h such that $\Phi^{\mu+h}(M) = \Phi^\mu(M)$ if $\mu \geq n$. This implies that the free resolution of M is periodic except for a finite part. ∎

Chapter 9. One-dimensional local rings of finite representation type

We have shown in the last chapter that Gorenstein rings of finite representation type are simple singularities. This chapter aims at showing that the converse is true for one-dimensional local rings. More generally we are able to provide a necessary and sufficient condition for one dimensional local rings to be of finite representation type; see Theorem (9.2). Furthermore we can draw the AR quivers for simple singularities of dimension one.

Throughout the chapter $(R, \mathfrak{m}, k)$ is a *one-dimensional* analytic local algebra over k, where k is an *algebraically closed field of characteristic* 0. Since we are interested in the finiteness of representation type, we always assume R has only an isolated singularity, or equivalently R is *reduced*; see (3.1) and (4.22). As before let $\mathfrak{C}(R)$ be the category of CM modules over R. Recall that the objects in $\mathfrak{C}(R)$ are exactly the modules without torsion, (1.5.2).

(9.1) DEFINITION. Let R^* be the integral closure of R in its total quotient ring. Note that R^* is also a one-dimensional (not necessarily local) ring which is finite over R. A local ring S is said to **birationally dominate** R if $R \subset S \subset R^*$.

One of the aims of this chapter is to prove

(9.2) THEOREM. (Greuel-Knörrer [32]) *The following two conditions are equivalent:*
(9.2.1) *R is of finite representation type;*
(9.2.2) *R birationally dominates a simple curve singularity.*

In particular we have:

(9.3) COROLLARY. *A local ring of simple singularity of dimension one is of finite representation type.*

For example the subring $k\{t^3, t^4, t^5\}$ of $k\{t\}$ birationally dominates $k\{t^3, t^4\}$, (E_6), and $k\{t^3, t^5\}$, (E_8), hence it is of finite representation type.

We start by remarking that finiteness of representation type will be inherited by birational dominance. More precisely,

(9.4) LEMMA. *Assume S birationally dominates R. If $\mathfrak{C}(R)$ is of finite representation type, then so is $\mathfrak{C}(S)$.*

PROOF: Let M be a CM module over S. Since $R \subset S$, M can be regarded as a module over R. It is then clear that M is torsion-free as an R-module, hence $M \in \mathfrak{C}(R)$. Let N be another CM module over S. Note that

(9.4.1) $$\text{Hom}_R(M, N) = \text{Hom}_S(M, N).$$

Indeed, $\text{Hom}_S(M, N)$ is naturally considered as a subset of $\text{Hom}_R(M, N)$. Here one can see that any element f in $\text{Hom}_R(M, N)$ is an S-homomorphism as follows: Let $s \in S$ and let $x \in M$. Since there is a nonzero divisor r in R with $rs \in R$, we have $rf(sx) = f(rsx) = rsf(x)$. Hence $f(sx) = sf(x)$, because r is a nonzero divisor on N. This shows that f is an S-homomorphism.

Now let M be an indecomposable CM module over S. Then it is also an indecomposable CM module as an R-module. In fact, $\text{End}_R(M) = \text{End}_S(M)$ by (9.4.1), and this is a (noncommutative) local ring. It also follows from (9.4.1) that if M and N are nonisomorphic indecomposable CM modules over S, then they are not isomorphic to each other when regarded as R-modules. Therefore the set of isomorphism classes of indecomposable CM modules over S is a subset of that over R. ∎

The following lemma is the key in the proof of one implication in the theorem.

(9.5) LEMMA. (Green-Reiner [31], Jacobinski [41]) *Let R be as above and suppose that R is of finite representation type. Then the following inequalities hold:*
(9.5.1) $length_R(R^*/\mathfrak{m}R^*) \leq 3$, and
(9.5.2) $length_R(\mathfrak{m}R^* + R/\mathfrak{m}^2 R^* + R) \leq 1.$

PROOF: (9.5.1): Write $T = R^*/\mathfrak{m}R^*$, $l = length_R(T)$ and denote by X the Grassmann space consisting of all two-dimensional k-subspaces of T. Note that X is an algebraic variety over k of dimension $2(l-2)$. Since R^* is a semi-local principal ideal ring, we see that T has only a finite number of ideals. Therefore we may find a (Zariski-)open set Y of X so that any elements in Y generate the same ideal in T. Let G be the group of units in T, which is an algebraic group of dimension at most l. It is easily seen that G acts on Y by multiplication of elements in T. Furthermore by this action the subgroup k^* consisting of all nonzero elements in k acts trivially, therefore the group G/k^* acts on Y.

For any element η in Y we denote by $L(\eta)$ the inverse image of η in R^*. Note that $L(\eta)$ $(\eta \in Y)$ are fractional R-ideals in R^*, hence they are CM modules over R, and that $L(\eta)R^* = L(\eta')R^*$ $(\eta, \eta' \in Y)$. Suppose $L(\eta) \simeq L(\eta')$ as R-modules for $\eta, \eta' \in Y$. Then, since $L(\eta)R^* = L(\eta')R^*$, the isomorphism is realized as a multiplication mapping by a unit element in R^*. Hence there is an element g in G such that $g(\eta) = \eta'$. This shows that there is a surjective mapping from the set of isomorphism classes of $L(\eta)$ ($\eta \in Y$) onto the orbit space of G/k^* in Y. Since R is of finite representation type, we conclude that the number of orbits is finite, therefore $\dim(Y) \leq \dim(G) - 1$, or $2(l-2) \leq l-1$. This exactly means $l \leq 3$.

Before commencing the proof of (9.5.2) we note the following:

(9.5.3) If there is an infinite set $\{S_\alpha| \alpha \in \Lambda\}$ of R-subalgebras of R^* with $S_\alpha \neq S_\beta$ for $\alpha \neq \beta$, then R is not of finite representation type.

To see this, first note that each S_α is a CM module when regarded as an R-module, since it is a submodule of R^* hence torsion-free. Next note that

(*) $$\operatorname{End}_R(S_\alpha) \simeq S_\alpha \quad \text{as an } R\text{-algebra.}$$

In fact, as in (9.4.1) we have $\operatorname{End}_R(S_\alpha) \simeq \operatorname{End}_{S_\alpha}(S_\alpha)$ and the latter is isomorphic to S_α. This proves (*). Now we show that $\{S_\alpha| \alpha \in \Lambda\}$ are all nonisomorphic R-modules. In fact, if $S_\alpha \simeq S_\beta$ as an R-module, then by (*) we see that $S_\alpha \simeq S_\beta$ as an R-algebra. Since S_α and S_β have the common total quotient ring, the last isomorphism gives the equality $S_\alpha = S_\beta$, hence $\alpha = \beta$ as required. If the total quotient ring of R is the product of n fields, then clearly each S_α is decomposed as a direct sum of at most n R-modules. Therefore there are infinitely many nonisomorphic indecomposable R-summands of S_α ($\alpha \in \Lambda$). Thus we have shown (9.5.3).

Now we prove (9.5.2). Assume the contrary. Then we can choose f and g in $\mathfrak{m} R^*$ so that their images in $\mathfrak{m} R^* + R/\mathfrak{m}^2 R^* + R$ are linearly independent over k. For any α in k, we put $S_\alpha = R[f + \alpha g]$ which is an R-subalgebra of R^*. Note that the Jacobson radical J_α of S_α is $(\mathfrak{m}, f + \alpha g)S_\alpha$. We claim that

(**) $$S_\alpha \neq S_\beta \quad \text{if} \quad \alpha \neq \beta.$$

To see this, let $\bar{J}_\alpha$ denote the R-module $J_\alpha/(\mathfrak{m}^2 R^* + R) \cap J_\alpha$ which is a submodule of $\mathfrak{m} R^* + R/\mathfrak{m}^2 R^* + R$. Since $\mathfrak{m}(f + \alpha g) \subset \mathfrak{m}^2 R^* + R$, it is easy to see that $\bar{J}_\alpha$ is a k-vector space generated by a single element $f + \alpha g$. If $S_\alpha = S_\beta$, then it is obvious that $J_\alpha = J_\beta$ hence that $\bar{J}_\alpha = \bar{J}_\beta$. This exactly means that $f + \alpha g$ and $f + \beta g$ generate the same subspace in $\mathfrak{m} R^* + R/\mathfrak{m}^2 R^* + R$, which forces $\alpha = \beta$, since f and g are linearly independent. This proves (**). Thus (9.5.2) follows by applying (9.5.3) to $\{S_\alpha| \alpha \in k\}$. ∎

(9.6) REMARK. It is known that the conditions (9.5.1) and (9.5.2) are necessary and sufficient for R to be of finite representation type. See Green-Reiner [31] for the details.

Recall that the **conductor** $\mathfrak{c}$ of R is an ideal of both R and R^*, which is defined to be $\{x \in R^*|\ xR^* \subset R\}$. Since $R^* \simeq k\{t\} \times k\{t\} \times \cdots \times k\{t\}$ (a product of m local rings for some m), we may consider the valuation v_i on the i-th component $k\{t\}$ and define $v = v_1 \times v_2 \times \cdots \times v_m : R^* \to \mathbf{N}_\infty^{(m)}$ where $\mathbf{N}_\infty = \mathbf{N} \cup \{\infty\}$. Then $v(R)$ is a subsemigroup of $\mathbf{N}_\infty^{(m)}$. Define the partial order on $\mathbf{N}_\infty^{(m)}$ as follows: $\underline{a} = (a_1, a_2, \ldots, a_m) \leq \underline{b} = (b_1, b_2, \ldots, b_m)$ if and only if $a_i \leq b_i$ for all i. We note that a k-subalgebra S of R^* is birational with R^* if and only if the set $\mathbf{N}_\infty^{(m)} - v(S)$ is finite. In particular the set $C = \{\underline{a} \in v(R)|\ \underline{b} \in v(R)$ for all $\underline{b} = (b_1, b_2, \ldots, b_m)$ with $\underline{a} \leq \underline{b}\}$ is nonempty. Define $\underline{c} = (c_1, c_2, \ldots, c_m)$ to be the (unique) minimal element in C. Then it is easy to see that the conductor ideal of R is generated by the element $(t^{c_1}, t^{c_2}, \ldots, t^{c_m})$ as an ideal of R^*. We call $\underline{c}$ **the conductor of the semigroup** $v(R)$.

Also $v(R) \cap \{\underline{a} \in \mathbf{N}_\infty^{(m)}|\ \underline{a} \gneq (0, 0, \ldots, 0)\}$ has a unique minimal element $(e_1, e_2, \ldots, e_m)$; it is not hard to see that $e = \sum_{i=1}^m e_i$ is the multiplicity of R along $\mathfrak{m}$.

The implication (9.2.1) $\Rightarrow$ (9.2.2) follows from the next lemma together with (9.5).

(9.7) LEMMA. *Suppose R satisfies the conditions (9.5.1) and (9.5.2). Then R birationally dominates a simple curve singularity.*

PROOF: First of all recall that $length_R(R^*/\mathfrak{m}R^*)$ is equal to the multiplicity e of the ring R along $\mathfrak{m}$. Therefore the condition (9.5.1) forces R to have m irreducible components with $m \leq 3$.

Suppose first that $m = 1$, and hence R is an integral domain and $R^* = k\{t\}$.

If $e = 2$, then R contains an element x with $v(x) = 2$. Then applying (8.6) to R^* we may assume $t^2 \in R$. Let n be the conductor of the semigroup $v(R)$. Notice that n is an even integer and that $t^{n+1} \in R$. Hence R birationally dominates a ring $k\{t^2, t^{n+1}\}$ that is a simple curve singularity of type (A_n).

If $e = 3$, then by (8.6) we may assume that $t^3 \in R$. Note that $\mathfrak{m}R^*/\mathfrak{m}^2 R^*$ has a k-basis $\{t^3, \alpha, \beta\}$ with $v(\alpha) = 4$ and $v(\beta) = 5$. Thus one sees from (9.5.2) that R contains γ with $v(\gamma) = 4$ or 5.

If $v(\gamma) = 4$, then the conductor of $v(R)$ is at most 6, since $\{3, 4\} \subset v(R)$. Hence we have $t^4 + at^5 \in R$ for some $a \in k$. In this case R birationally dominates the ring $k\{t^3, t^4 + at^5\} \simeq k\{y, x\}/(f)$ where $f = x^3 - 3axy^3 - y^4(1 + a^3 y)$. Then by $(iii - 2)$ from the proof of (8.5), f can be changed into the form $x^3 + y^4$ after applying a suitable automorphism of $k\{y, x\}$. Therefore R birationally dominates the simple singularity of type (E_6).

If $v(\gamma) = 5$, then the conductor of $v(R)$ is at most 8, since $\{3, 5\} \subset v(R)$. Hence we have $t^5 + at^7 \in R$ for some $a \in k$. Then R conatins the ring $k\{t^3, t^5 + at^7\} \simeq k\{y, x\}/(f)$ where $f = x^3 - 3axy^4 - y^5(1 + a^3 y^2)$. Here applying again $(iii - 2)$ from the proof of

(8.5), we can change f into the form $x^3 + y^5$, whence R birationally dominates the simple singularity of type (E_8).

Now suppose that $m = 2$ and that $R^* = k\{t\} \times k\{t\}$.

If $e = 2$, then by (8.6) we may assume that (t,t) belongs to R. Let (c_1, c_2) be the conductor of $v(R)$ and take an integer n with $n \geq max\{c_1, c_2\}$. Then from the definition of conductors we see that $(t^n, -t^n) \in R$, therefore R birationally dominates the ring $k\{(t,t), (t^n, -t^n)\} \simeq k\{x,y\}/(y^2 - x^{2n})$ that is the simple singularity of type (A_{2n-1}).

If $e = 3$, then we may assume that $(1,2) \in v(R)$. Consider the semigroup $H = \{h \in \mathbb{N}_\infty | (\infty, h) \in v(R)\}$. First we claim that $\{2, 3\} \cap H \neq \emptyset$. Indeed, if not, there would be no element in R with the value (∞, μ) for $\mu = 2, 3$. Since $v(\mathfrak{m}R^*) = [1, \infty] \times [2, \infty]$, this would imply that the module $\mathfrak{m}R^* + R/\mathfrak{m}^2 R^* + R$ has linearly independent elements $\bar{x}$ and $\bar{y}$ whose inverse images in R^* have values $v(x) = (\infty, 2)$, $v(y) = (\infty, 3)$. This contradicts the condition (9.5.2), therefore either 2 or 3 belongs to H.

If $2 \in H$, then there is an element $(0, \alpha)$ with $v_2(\alpha) = 2$, thus making use of (8.6) we may assume that $\xi = (0, t^2) \in R$. Denoting by n the conductor of the semigoup H, we have $\eta = (t, t^2 + c_m t^m + c_{m+1} t^{m+1} + \ldots + c_{n-1} t^{n-1}) \in R$ for some $c_j \in k$ ($m \leq j \leq n-1$) with $c_m \neq 0$, because $(1,2) \in v(R)$. Note that n is an even integer and that m can be taken as an odd integer. Subtracting a multiple of ξ from η and using the fact that any $(0, t^l)$ ($l \geq n$) is in R, we may assume that $(t, t^m) \in R$. Thus R birationally dominates the ring $k\{(t, t^m), (0, t^2)\} \simeq k\{x,y\}/(y(x^2 - y^m))$, the simple singularity of type (D_{m+2}).

Suppose $3 \in H$. We may assume that $(t, t^2) \in R$. Letting $(0, \alpha)$ be an element with $v_2(\alpha) = 3$, we see that R contains $(0, \alpha t^2)$, $(0, \alpha^2)$ and $(0, \alpha t^4)$ whose values are $(\infty, 5)$, $(\infty, 6)$ and $(\infty, 7)$, hence H has conductor at most 5. Thus $(0, t^3 + at^4) \in R$ for some $a \in k$. Then R birationally dominates the ring $k\{(t, t^2), (0, t^3 + at^4)\} \simeq k\{y, x\}/(x(x^2 - 2axy^2 - y^3 + a^2 y^4))$ which is the ring of simple singularity of type (E_7) by $(iii - 1)$ from the proof of (8.5).

Finally consider the case when $m = 3$ and hence $R^* = k\{t\} \times k\{t\} \times k\{t\}$.

In this case we have $e = 3$ and $(1,1,1) \in v(R)$. We claim that one of $(1, 1, \infty)$, $(1, \infty, 1)$ and $(\infty, 1, 1)$ belongs to $v(R)$. Indeed, if none of them were in $v(R)$, then the elements α and β in R^* with $v(\alpha) = (1, 1, \infty)$, $v(\beta) = (1, \infty, 1)$ would give linearly independent elements in $\mathfrak{m}R^* + R/\mathfrak{m}^2 R^* + R$, which would be against the condition (9.5.2). Thus we may assume that $(1, 1, \infty) \in v(R)$, and hence using (8.6) we may also assume that $y = (t, t, 0) \in R$. Since there is an element x with $v(x) = (1, 1, 1)$, applying (8.6) to the third component and subtracting a suitable power series of y from x, we can take $x = (t + p(t), t, t) \in R$ for some $p(t)$ with $v_1(p(t)) \geq 2$. Then R birationally dominates the ring $S = k\{x, y\}$ with the relation $f = y(x - y)(x - y - p(y))$. Since $\{f = 0\}$ has two or three different tangent directions and S has multiplicity 3, we see by (ii) from the proof of (8.5) that S is a simple curve singularity of D-type. ∎

Proof of (9.2.2) ⇒ (9.2.1): We provide below two different proofs of this.

For the first, we assume some results to be proved in later chapters. (Since they will be proved independently of this chapter, there is no fear of a logical cycle.)

Let R be a local ring which birationally dominates a simple singularity S. By (9.4) it suffices to show that $\mathfrak{C}(S)$ is of finite representation type. Let $S = k\{x,y\}/(f)$ with f one of the polynomials in (8.5). Define then a simple singularity $S^\sharp$ of dimension two as follows:

$$S^\sharp = k\{x,y,z\}/(f+z^2)$$

In the next chapter we will see that simple singularities of dimension two are always of finite representation type; see (10.14) and (10.15). On the other hand, it will be seen in Chapter 12 that the finiteness of representation type of S is equivalent to that of $S^\sharp$ (Theorem (12.5)). From these results we obtain that $\mathfrak{C}(S)$ is of finite representation type and the proof is completed.

The second proof of this is more direct but also more complicated. As above it is sufficient to show that simple singularities are of finite representation type. Therefore if we have a complete description of AR quivers for simple singularities and if we know they are finite, the proof will be completed. We shall describe the AR quivers for simple curve singularities below to check their finiteness.

First notice the following fact.

(9.8) LEMMA. *Let R be an analytic reduced local ring that is a hypersurface of dimension one. Then the AR translation τ is given by*

(9.8.1) $$\tau(M) \simeq \mathrm{syz}^1_R M \qquad (M \in \mathfrak{C}(R)),$$

and τ satisfies $\tau^2 = 1$. Furthermore, if $\mathfrak{m} = \sum_i M_i$ is a decomposition of $\mathfrak{m}$ into indecomposable modules, then the natural inclusions $M_i \to R$ are the all of irreducible morphisms ending in R. Likewise, there are irreducible morphisms $R \to \tau(M_i)$ starting from R.

PROOF: Let $\cdots \to F_1 \to F_0 \to M \to 0$ be a free resolution of M. Then by definition we have an exact sequence $0 \to M^* \to F_0^* \to F_1^* \to \mathrm{tr}(M) \to 0$, and hence $0 \to M^* \to F_0^* \to (\tau(M))^* \to 0$ by (3.11). Since R is a Gorenstein ring, using (1.13) we have an exact sequence $0 \to \tau(M) \to F_0 \to M \to 0$, and (9.8.1) follows from this.

Since any CM module over R has a periodic free resolution with periodicity 2 (c.f.(7.2)), we see that $\mathrm{syz}^2 M \simeq M$ for any $M \in \mathfrak{C}(R)$, in particular, $\tau^2(M) = M$.

Let $X \to R$ be an irreducible morphism. Then the image lies inside the maximal ideal, and hence the map is decomposed as $X \to \mathfrak{m} \subset R$. Therefore by definition of irreducible morphisms, $X \simeq M_i$ for some i and the morphism is a natural inclusion. This proves the second statement. Furthermore by the duality theorem, $R \to M_i^*$ are irreducible morphims starting from R. Since R is an isolated singularity, there is an AR sequence starting from M_i that must be of the form:

$$0 \to M_i \to R \oplus M \to M_i^* \to 0,$$

for some $M \in \mathfrak{C}(R)$. Therefore $\tau(M_i^*) = M_i$, and as a consequence $\tau(M_i) = \tau^2(M_i^*) = M_i^*$. ∎

Let us start drawing the AR quivers.

The singularities of type (A_n) with n even were done in (5.12). So we begin with rings of type (A_n) for odd n.

(9.9) Let $f = x^2 + y^{n+1}$ and let $R = k\{x, y\}/(f)$ where n is an odd integer. Since $f = (y^{(n+1)/2} + ix)(y^{(n+1)/2} - ix)$ with $i = \sqrt{-1}$, the modules $N_\pm := R/(y^{(n+1)/2} \pm ix)$ are CM modules over R. N_+ (resp. N_-) is given by the matrix factorization $(y^{(n+1)/2} + ix, y^{(n+1)/2} - ix)$ (resp. $(y^{(n+1)/2} - ix, y^{(n+1)/2} + ix)$). Consider the square matrices

$$\varphi_j = \begin{pmatrix} x & y^j \\ y^{n+1-j} & -x \end{pmatrix} \qquad (0 \leq j \leq n+1)$$

on $k\{x, y\}$. It is clear that each (φ_j, φ_j) gives a matrix factorization of f. Let $M_j = \mathrm{Coker}(\varphi_j, \varphi_j)$. It is obvious that $M_0 \simeq R$, $M_j \simeq M_{n+1-j}$ and $M_{(n+1)/2} \simeq N_+ \oplus N_-$.

First of all we show that $N_\pm$ and M_j ($0 \leq j \leq (n-1)/2$) are all indecomposable. For $N_\pm$ and M_0 this is clear because they are generated by single elements. It is true as well that any CM module generated by one element is isomorphic to one of them. If M_j ($1 \leq j \leq (n-1)/2$) were decomposed, say $M_j \simeq A \oplus B$, then A and B would be generated by one element and nonfree, and so isomorphic to $N_\pm$. Since $I(\varphi_j) = (x, y^j)$ is not equal to any of $I(N_+)$, $I(N_-)$ and $I(N_+) + I(N_-)$, this is a contradiction. (See (8.11) for the definition of I.)

Next it follows from (9.8) and (7.7) that

$$\tau(M_j) = M_j, \quad \tau(N_+) = N_- \quad \text{and} \quad \tau(N_-) = N_+.$$

Since we have an exact sequence $0 \to N_- \to R \to N_+ \to 0$, we obtain an epimorphism $\mathrm{End}_R(N_+) \to \mathrm{Ext}^1_R(N_+, N_-)$ and it is easy to see that the endomorphism given by multiplication by $y^{(n-1)/2}$ on N_+ is sent to the socle element of $\mathrm{Ext}^1_R(N_+, N_-)$. Thus the exact sequence corresponding to the socle of $\text{Ext}^1_R(N_+, N_-)$ is given by the extension

$$\begin{pmatrix} y^{(n+1)/2} - ix & y^{(n-1)/2} \\ 0 & y^{(n+1)/2} + ix \end{pmatrix}.$$

Hence the AR sequence ending in N_+ is

$$0 \longrightarrow N_- \longrightarrow L \longrightarrow N_+ \longrightarrow 0;$$

see (3.13) and (7.8), where one can easily see that L is isomorphic to $M_{(n-1)/2}$. Similarly we can obtain the AR sequence ending N_-:

$$0 \longrightarrow N_+ \longrightarrow M_{(n-1)/2} \longrightarrow N_- \longrightarrow 0.$$

In the same way as this the extensions

$$\begin{pmatrix} \varphi_j & \varepsilon_j \\ 0 & \varphi_j \end{pmatrix} \quad \text{with} \quad \varepsilon_j = \begin{pmatrix} 0 & y^{j-1} \\ -y^{n-j} & 0 \end{pmatrix}$$

give the AR sequences

$$0 \longrightarrow M_j \longrightarrow M_{j-1} \oplus M_{j+1} \longrightarrow M_j \longrightarrow 0,$$

for all j $(1 \leq j \leq (n-1)/2)$. Consequently we obtain a part of the AR quiver for R:

Figure (9.9.1). (A_n) for odd n

We show that this graph is a connected component of the AR quiver. For this it is sufficient to show that if $n \geq 3$ (resp. $n = 1$), then there is no CM module other than M_1 (resp. N_+ and N_-) that is connected to R by arrows. It is, however, a direct consequence of (9.8), since $M_1 \simeq \mathfrak{m}$ if $n \geq 3$ and $N_+ \oplus N_- \simeq \mathfrak{m}$ if $n = 1$. Thus the graph (9.9.1) is the connected component of the AR quiver of R, and Theorem (6.2) implies that this is the whole quiver.

(9.10) *Exercise.* Prove (5.11) again by using the same argument as above.

(9.11) Next consider the simple singularities of type (D_n) where n is an *odd* integer. Let $R = k\{x, y\}/(f)$ where $f = x^2 y + y^{n-1}$ with n odd. Then $(\alpha, \beta) = (y, x^2 + y^{n-2})$ and $(\beta, \alpha) = (x^2 + y^{n-2}, y)$ are matrix factorizations of f. We denote $\mathrm{Coker}(\alpha, \beta)$ (resp. $\mathrm{Coker}(\beta, \alpha)$) by A (resp. B).

We take the 2×2 matrices:

(9.11.1) $$\varphi_j = \begin{pmatrix} x & y^j \\ y^{n-j-2} & -x \end{pmatrix}, \qquad \psi_j = \begin{pmatrix} xy & y^{j+1} \\ y^{n-j-1} & -xy \end{pmatrix},$$

(9.11.2) $$\xi_j = \begin{pmatrix} x & y^j \\ y^{n-j-1} & -xy \end{pmatrix}, \qquad \eta_j = \begin{pmatrix} xy & y^j \\ y^{n-j-1} & -x \end{pmatrix},$$

for j with $0 \leq j \leq n-3$. It can be easily seen that (φ_j, ψ_j) and (ξ_j, η_j) are matrix factorizations of f. Putting

(9.11.3) $$M_j = \mathrm{Coker}(\varphi_j, \psi_j), \qquad N_j = \mathrm{Coker}(\psi_j, \varphi_j),$$
$$X_j = \mathrm{Coker}(\xi_j, \eta_j), \qquad Y_j = \mathrm{Coker}(\eta_j, \xi_j),$$

we see that

(9.11.4) $$M_0 \simeq B, \quad N_0 \simeq A \oplus R, \quad X_0 \simeq R, \quad Y_0 \simeq R,$$

and that

$$X_{(n-1)/2} \simeq Y_{(n-1)/2}.$$

Furthermore,

(9.11.5) $$M_j \simeq M_{n-j-2}, \quad N_j \simeq N_{n-j-2}, \quad X_j \simeq Y_{n-j-1} \quad \text{and} \quad Y_j \simeq X_{n-j-1},$$

for $1 \leq j \leq n-3$. The modules in (9.11.3) are indecomposable CM modules whenever $j \geq 1$. Notice that M_j (resp. Y_j) is isomorphic to the ideal generated by (xy, y^{j+1}) (resp. (x, y^j)).

The AR sequences are given by the following extensions:

$$\begin{pmatrix} \alpha & x \\ 0 & \beta \end{pmatrix} \quad : 0 \to A \to X_1 \to B \to 0,$$

$$\begin{pmatrix} \beta & x \\ 0 & \alpha \end{pmatrix} \quad : 0 \to B \to Y_1 \to A \to 0,$$

$$\begin{pmatrix} \varphi_j & \varepsilon_j \\ 0 & \psi_j \end{pmatrix} \quad : 0 \to M_j \to X_j \oplus Y_{j+1} \to N_j \to 0,$$

$$\begin{pmatrix} \psi_j & \varepsilon_j \\ 0 & \varphi_j \end{pmatrix} \quad : 0 \to N_j \to X_{j+1} \oplus Y_j \to M_j \to 0,$$

$$\begin{pmatrix} \xi_j & \varepsilon_{j-1} \\ 0 & \eta_j \end{pmatrix} \quad : 0 \to X_j \to M_{j-1} \oplus N_j \to Y_j \to 0,$$

$$\begin{pmatrix} \eta_j & \varepsilon_j \\ 0 & \xi_j \end{pmatrix} \quad : 0 \to Y_j \to M_j \oplus N_{j-1} \to X_j \to 0,$$

where $\varepsilon_j = \begin{pmatrix} 0 & y^j \\ -y^{n-j-2} & 0 \end{pmatrix}$. Therefore one of the connected components of the AR quiver of R is obtained:

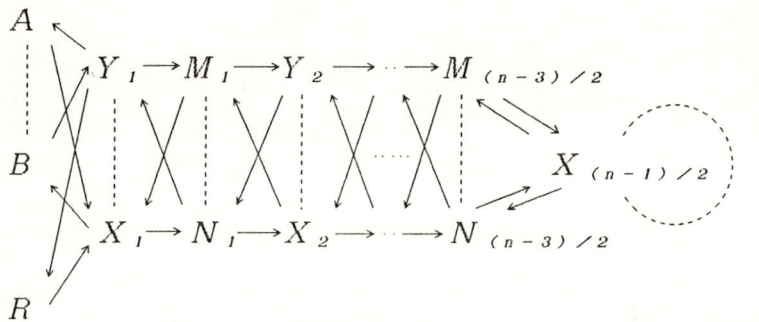

Figure (9.11.5). (D_n) for odd n

Using (9.8) and (6.2) we may claim that this graph is the whole quiver and that the ring R must be of finite representation type.

(9.12) Now consider the simple curve singularity of type (D_n) with n even. Let $R = k\{x,y\}/(f)$ where $f = x^2y + y^{n-1}$ with n even. In this case the AR quiver will be obtained by a subtle change of (9.11).

Let $A = \text{Coker}(y, x^2 + y^{n-2})$, $B = \text{Coker}(x^2 + y^{n-2}, y)$ as in (9.11) and define matrices and modules as in (9.11.1), (9.11.2) and (9.11.3). Furthermore define modules as follows:

$$C_+ = \text{Coker}(y(x + iy^{(n-2)/2}), x - iy^{(n-2)/2}),$$
$$D_+ = \text{Coker}(x - iy^{(n-2)/2}, y(x + iy^{(n-2)/2})),$$
$$C_- = \text{Coker}(y(x - iy^{(n-2)/2}), x + iy^{(n-2)/2}),$$
$$D_- = \text{Coker}(x + iy^{(n-2)/2}, y(x - iy^{(n-2)/2})).$$

The equalities in (9.11.4) and (9.11.5) are still valid and

$$M_{(n-2)/2} \simeq D_+ \oplus D_-, \qquad N_{(n-2)/2} \simeq C_+ \oplus C_-.$$

The AR sequences are obtained by the same extensions as in (9.11) and the AR quiver is:

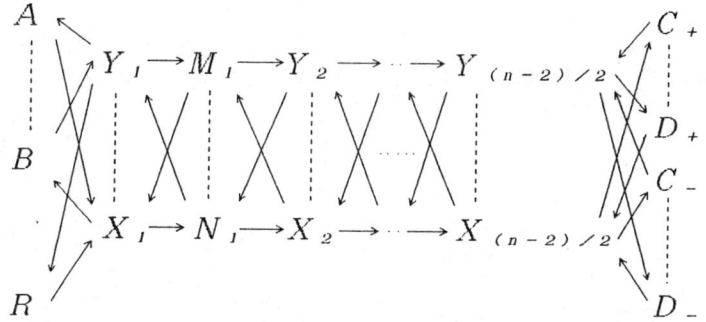

Figure (9.12.1). (D_n) for even n

(9.13) Now let us consider the case (E_6) and so let $R = k\{x,y\}/(f)$ with $f = x^3 + y^4$. Take the matrices

$$\varphi_1 = \begin{pmatrix} x & y \\ y^3 & -x^2 \end{pmatrix}, \qquad \psi_1 = \begin{pmatrix} x^2 & y \\ y^3 & -x \end{pmatrix},$$

$$\varphi_2 = \begin{pmatrix} x & y^2 \\ y^2 & -x^2 \end{pmatrix}, \qquad \psi_2 = \begin{pmatrix} x^2 & y^2 \\ y^2 & -x \end{pmatrix},$$

$$\alpha = \begin{pmatrix} y^3 & x^2 & xy^2 \\ xy & -y^2 & x^2 \\ x^2 & -xy & -y^3 \end{pmatrix}, \qquad \beta = \begin{pmatrix} y & 0 & x \\ x & -y^2 & 0 \\ 0 & x & -y \end{pmatrix},$$

each pair of which gives a matrix factorization of f. We define the CM modules as follows:

$$M_i = \text{Coker}(\varphi_i, \psi_i), \quad N_i = \text{Coker}(\psi_i, \varphi_i) \quad (i = 1, 2),$$
$$A = \text{Coker}(\alpha, \beta), \quad B = \text{Coker}(\beta, \alpha).$$

It is easy to see that there are isomorphisms

$$N_1 \simeq \mathfrak{m}, \quad M_1 \simeq (x^2, y)R, \quad N_2 \simeq M_2 \simeq (x^2, y^2)R, \quad B \simeq (x^2, xy, y^2)R,$$

so that they are ideals, and A has rank two. Furthermore,

$$\tau(M_1) = N_1, \quad \tau(N_1) = M_1, \quad \tau(M_2) = M_2, \quad \tau(A) = B \quad \text{and} \quad \tau(B) = A.$$

The extensions

$$\begin{pmatrix} \varphi_1 & \varepsilon_1 \\ 0 & \psi_1 \end{pmatrix}, \quad \begin{pmatrix} \psi_1 & \varepsilon_2 \\ 0 & \varphi_1 \end{pmatrix},$$

where $\varepsilon_1 = \begin{pmatrix} 0 & 1 \\ -xy^2 & 0 \end{pmatrix}$ and $\varepsilon_2 = \begin{pmatrix} 0 & x \\ -y^2 & 0 \end{pmatrix}$, give the AR sequences

$$0 \longrightarrow M_1 \longrightarrow A \longrightarrow N_1 \longrightarrow 0,$$

$$0 \longrightarrow N_1 \longrightarrow B \oplus R \longrightarrow M_1 \longrightarrow 0.$$

On the other hand, the matrices

$$\xi = \begin{pmatrix} \varphi_2 & \varepsilon_3 \\ 0 & \psi_2 \end{pmatrix}, \quad \eta = \begin{pmatrix} \psi_2 & \varepsilon_4 \\ 0 & \varphi_2 \end{pmatrix},$$

with $\varepsilon_3 = \begin{pmatrix} 0 & y \\ -xy & 0 \end{pmatrix}$ and $\varepsilon_4 = \begin{pmatrix} 0 & xy \\ y & 0 \end{pmatrix}$, give a new indecomposable matrix factorization (ξ, η), and letting $X = \text{Coker}(\xi, \eta) \simeq \text{Coker}(\eta, \xi)$, we have the AR sequence

$$0 \longrightarrow M_2 \longrightarrow X \longrightarrow M_2 \longrightarrow 0.$$

The AR quiver is shown in Figure (9.13.1).

(9.14) Let $R = k\{x, y\}/(f)$ with $f = x^3 + xy^3$, and so R is a simple singularity of type (E_7). In this case the following pairs of matrices give (non-isomorphic) indecomposable

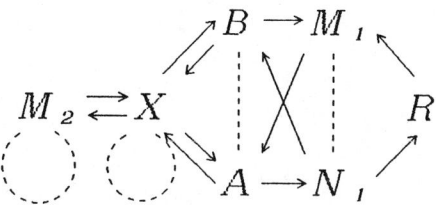

Figure (9.13.1). (E_6)

matrix-factorizations of f:

$$\alpha = (x), \qquad \beta = (x^2 + y^3),$$

$$\gamma = x\begin{pmatrix} x & y \\ y^2 & -x \end{pmatrix}, \qquad \delta = \begin{pmatrix} x & y \\ y^2 & -x \end{pmatrix},$$

$$\varphi_1 = \begin{pmatrix} x & y \\ xy^2 & -x^2 \end{pmatrix}, \qquad \psi_1 = \begin{pmatrix} x^2 & y \\ xy^2 & -x \end{pmatrix},$$

$$\varphi_2 = \begin{pmatrix} x & y^2 \\ xy & -x^2 \end{pmatrix}, \qquad \psi_2 = \begin{pmatrix} x^2 & y^2 \\ xy & -x \end{pmatrix},$$

$$\xi_1 = \begin{pmatrix} xy^2 & -x^2 & -x^2y \\ xy & y^2 & -x^2 \\ x^2 & xy & xy^2 \end{pmatrix}, \qquad \eta_1 = \begin{pmatrix} y & 0 & x \\ -x & xy & 0 \\ 0 & -x & y \end{pmatrix},$$

$$\xi_2 = \begin{pmatrix} x^2 & -y^2 & -xy \\ xy & x & -y^2 \\ xy^2 & xy & x^2 \end{pmatrix}, \qquad \eta_2 = \begin{pmatrix} x & 0 & y \\ -xy & x^2 & 0 \\ 0 & -xy & x \end{pmatrix},$$

and finally,

$$\xi_3 = \begin{pmatrix} \gamma & \varepsilon \\ 0 & \delta \end{pmatrix}, \qquad \eta_3 = \begin{pmatrix} \delta & -\varepsilon \\ 0 & \gamma \end{pmatrix},$$

with $\varepsilon = \begin{pmatrix} y & 0 \\ 0 & y \end{pmatrix}$. Now letting

$$A = \mathrm{Coker}(\alpha, \beta), \quad B = \mathrm{Coker}(\beta, \alpha), \quad C = \mathrm{Coker}(\gamma, \delta), \quad D = \mathrm{Coker}(\delta, \gamma),$$
$$M_i = \mathrm{Coker}(\varphi_i, \psi_i), \quad N_i = \mathrm{Coker}(\psi_i, \varphi_i), \quad X_i = \mathrm{Coker}(\xi_i, \eta_i), \quad Y_i = \mathrm{Coker}(\eta_i, \xi_i),$$

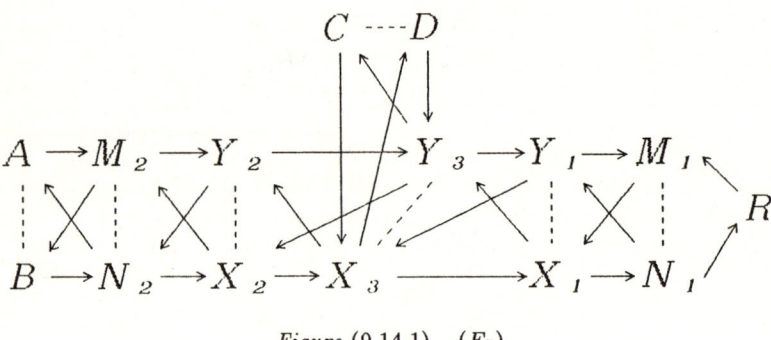

Figure (9.14.1). (E_7)

for $i = 1, 2$, we obtain the AR quiver as shown in Figure (9.14.1).

(9.15) Let R be a simple curve singularity of type (E_8), so $R = k\{x,y\}/(f)$ with $f = x^3 + y^5$. Then the indecomposable matrix-factorizations of f are one of the following pairs of matrices:

$$\varphi_1 = \begin{pmatrix} x & y \\ y^4 & -x^2 \end{pmatrix}, \qquad \psi_1 = \begin{pmatrix} x^2 & y \\ y^4 & -x \end{pmatrix}$$

$$\varphi_2 = \begin{pmatrix} x & y^2 \\ y^3 & -x^2 \end{pmatrix}, \qquad \psi_2 = \begin{pmatrix} x^2 & y^2 \\ y^3 & -x \end{pmatrix}$$

$$\alpha_1 = \begin{pmatrix} y & -x & 0 \\ 0 & y & -x \\ x & 0 & y^3 \end{pmatrix}, \qquad \beta_1 = \begin{pmatrix} y^4 & xy^3 & x^2 \\ -x^2 & y^4 & xy \\ -xy & -x^2 & y^2 \end{pmatrix}$$

$$\alpha_2 = \begin{pmatrix} y & -x & 0 \\ 0 & y^2 & -x \\ x & 0 & y^2 \end{pmatrix}, \qquad \beta_2 = \begin{pmatrix} y^4 & xy^2 & x^2 \\ -x^2 & y^3 & xy \\ -xy^2 & -x^2 & y^3 \end{pmatrix}$$

$$\gamma_1 = \begin{pmatrix} y & -x & 0 & y^3 \\ x & 0 & -y^3 & 0 \\ -y^2 & 0 & -x^2 & 0 \\ 0 & -y^2 & -xy & -x^2 \end{pmatrix}, \qquad \delta_1 = \begin{pmatrix} 0 & x^2 & -y^3 & 0 \\ -x^2 & xy & 0 & -y^3 \\ 0 & -y^2 & -x & 0 \\ y^2 & 0 & y & -x \end{pmatrix}$$

$$\gamma_2 = \begin{pmatrix} x & y^2 & 0 & y \\ y^3 & -x^2 & -xy^2 & 0 \\ 0 & 0 & x^2 & y^2 \\ 0 & 0 & y^3 & -x \end{pmatrix}, \qquad \delta_2 = \begin{pmatrix} x^2 & y^2 & 0 & xy \\ y^3 & -x & -y^2 & 0 \\ 0 & 0 & x & y^2 \\ 0 & 0 & y^3 & -x^2 \end{pmatrix}$$

One-dimensional local rings of finite representation type

$$\xi_1 = \begin{pmatrix} y^4 & xy^2 & x^2 & 0 & 0 & xy \\ -x^2 & y^3 & xy & -x & 0 & 0 \\ -xy^2 & -x^2 & y^3 & 0 & -xy & 0 \\ 0 & 0 & 0 & y & -x & 0 \\ 0 & 0 & 0 & 0 & y^2 & -x \\ 0 & 0 & 0 & x & 0 & y^2 \end{pmatrix}, \quad \eta_1 = \begin{pmatrix} y & -x & 0 & 0 & 0 & -x \\ 0 & y^2 & -x & xy & 0 & 0 \\ x & 0 & y^2 & 0 & xy & 0 \\ 0 & 0 & 0 & y^4 & xy^2 & x^2 \\ 0 & 0 & & -x^2 & y^3 & xy \\ 0 & 0 & 0 & -xy^2 & -x^2 & y^3 \end{pmatrix},$$

$$\xi_2 = \begin{pmatrix} y^4 & x^2 & 0 & -xy^2 & 0 \\ -x^2 & xy & 0 & -y^3 & 0 \\ 0 & -y^2 & -x & 0 & y^3 \\ -xy^2 & y^3 & 0 & x^2 & 0 \\ -y^3 & 0 & -y^2 & xy & -x^2 \end{pmatrix}, \quad \eta_2 = \begin{pmatrix} y & -x & 0 & 0 & 0 \\ x & 0 & 0 & y^2 & 0 \\ -y^2 & 0 & -x^2 & 0 & -y^3 \\ 0 & -y^2 & 0 & x & 0 \\ 0 & 0 & y^2 & y & -x \end{pmatrix}.$$

Defining the CM modules by

$$M_i = \operatorname{Coker}(\varphi_i, \psi_i), \quad N_i = \operatorname{Coker}(\psi_i, \varphi_i),$$
$$A_i = \operatorname{Coker}(\alpha_i, \beta_i), \quad B_i = \operatorname{Coker}(\beta_i, \alpha_i),$$
$$C_i = \operatorname{Coker}(\gamma_i, \delta_i), \quad D_i = \operatorname{Coker}(\delta_i, \gamma_i),$$
$$X_i = \operatorname{Coker}(\xi_i, \eta_i), \quad Y_i = \operatorname{Coker}(\eta_i, \xi_i) \qquad (i = 1, 2),$$

we can describe the AR quiver:

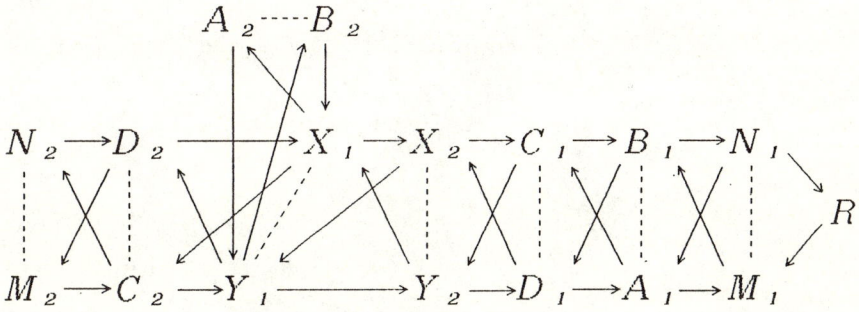

Figure (9.15.1). (E_8)

(9.16) REMARK. Once we know the AR quivers of simple curve singularities, it is easy to describe AR quivers for rings that birationally dominate simple singularities. For example, Figures (9.16.1) and (9.16.2) show the AR quivers of rings $k\{t^3, t^4, t^5\}$ and $k\{x, y, z\}/(xy, yz, zx)$ respectively.

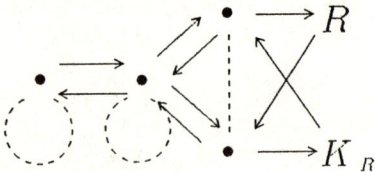

Figure (9.16.1)

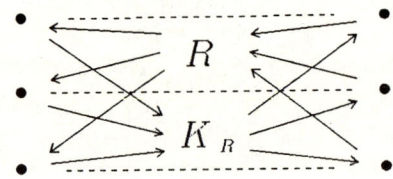

Figure (9.16.2)

Chapter 10. McKay graphs

We introduce below the McKay graph of a finite subgroup G in $\mathrm{GL}(2,k)$ and show that it is exactly the same as the AR quiver for the invariant subring of $k\{x,y\}$ by the natural action of G; see Theorem (10.14). All the materials in this chapter are taken from Auslander [6].

Let us begin with remarking on invariants by a finite group. In what follows, k is an *algebraically closed field* of characteristic 0 and G is a finite subgroup of $\mathrm{GL}(2,k)$. We denote by V a two dimensional k vector space with basis $\{x,y\}$ on which G naturally acts. Then the action of G can be extended to the ring $S = k\{x,y\}$, i.e., $\sigma \in G$ sends $f(x,y) \in S$ to $f(\sigma(x), \sigma(y))$. We now consider the invariant subring:

$$R = S^G = \{z \in S|\ \sigma(z) = z \text{ for all } \sigma \in G\}.$$

Denote the skew group ring by $S * G$, i.e., $S * G = \sum\{S\sigma|\ \sigma \in G\}$ as an S-module and the product is defined by

(10.1.1) $\qquad (s_1\sigma_1)(s_2\sigma_2) = s_1\sigma_1(s_2)\sigma_1\sigma_2 \quad (s_i \in S, \sigma_i \in G).$

An $S * G$-module M is exactly an S-module with G-action such that $\sigma(sm) = \sigma(s)\sigma(m)$ ($s \in S, m \in M, \sigma \in G$). Note that $f : M \to N$ is an $S * G$-homomorphism if and only if f is a G-homomorphism as well as an S-homomorphism. When M and N are $S * G$-modules, $\mathrm{Hom}_S(M, N)$ has the structure of $S * G$-module with G-action defined by

$$(\sigma f)(m) = \sigma f(\sigma^{-1} m) \quad (\sigma \in G, m \in M, f \in \mathrm{Hom}_S(M, N)).$$

Note that an element in $\mathrm{Hom}_S(M, N)$ is G-invariant only when it is an $S * G$-homomorphism. Hence,

$$\mathrm{Hom}_{S*G}(M, N) = \mathrm{Hom}_S(M, N)^G.$$

Since taking G-invariants is an exact functor, the derived functors of $\text{Hom}_{S*G}(\ ,\ )$ are obtained as follows:

$$\text{Ext}^i_{S*G}(M, N) = \text{Ext}^i_S(M, N)^G \quad (i \geq 0).$$

It easily follows from this that an $S*G$-module M is projective if and only if it is projective as an S-module.

Let kG be the group ring in usual sense and denote by $\mathfrak{M}(kG)$ the category of finitely generated left kG-modules. Denoting as well the category of finitely generated left $S*G$-modules by $\mathfrak{M}(S*G)$, we define the functor F from $\mathfrak{M}(kG)$ to $\mathfrak{M}(S*G)$ as follows:

(10.1.2) $$F(W) = S \otimes_k W, \qquad F(f) = 1_S \otimes_k f,$$

for a kG-module W and a kG-homomorphism f. Here the action of $S*G$ is given by

$$(s\sigma)(t \otimes w) = s\sigma(t) \otimes \sigma(w), \quad (s\sigma)(1_S \otimes f) = s \otimes \sigma(f) \quad (s, t \in S, \sigma \in G, w \in W).$$

We also denote by $\wp(S*G)$ the full subcategory of $\mathfrak{M}(S*G)$ consisting of all projective modules. We now have an elementary relation among these categories.

(10.1) **LEMMA.** *The functor F gives rise to a functor from $\mathfrak{M}(kG)$ into $\wp(S*G)$ which has the left adjoint functor $F' = S/\mathfrak{n} \otimes_S$ where $\mathfrak{n}$ is the maximal ideal $(x, y)S$ of S. Furthermore the functor gives a one-to-one correspondence between the sets of isomorpshim classes of objects in these two categories.*

PROOF: Let W be a kG-module. Clearly $F(W)$ is a free modules when regarded as S-module, therefore $F(W)$ is projective as an $S*G$-module. Hence F defines a functor from $\mathfrak{M}(kG)$ to $\wp(S*G)$ which we also denote by the same letter F.

It is clear from the definition that $F' \cdot F$ is the identity on $\mathfrak{M}(kG)$. We show that the composition $F \cdot F'$ gives the identity mapping on the objects in $\wp(S*G)$. To do this, let M be an arbitrary projective $S*G$-module. Recall that $F \cdot F'(M) = S \otimes_k (M/\mathfrak{n}M)$ and that M is free when regarded as an S-module. Note that the natural mapping $\pi : M \to M/\mathfrak{n}M$ gives the minimal projective cover as an $S*G$-module. In other words,

(*) \qquad for any proper submodule N of M, we have $\pi(N) \neq M/\mathfrak{n}M$.

In fact, if $\pi(N) = M/\mathfrak{n}M$ then $M = N + \mathfrak{n}M$, hence $M = N$ by Nakayama's lemma. Now consider a natural mapping $S \otimes_k (M/\mathfrak{n}M) \to M/\mathfrak{n}M$. Since this is also a projective cover as an $S*G$-module, there is an $S*G$-homomorphism which makes the following diagram commutative:

$$\begin{array}{ccccc} S \otimes_k (M/\mathfrak{n}M) & \longrightarrow & M/\mathfrak{n}M & \longrightarrow & 0 \\ {\scriptstyle f}\downarrow & & \| & & \\ M & \longrightarrow & M/\mathfrak{n}M & \longrightarrow & 0 \end{array}$$

We know from (*) that f must be surjective. Since M is projective, this shows that M is a summand of $S \otimes_k (M/\mathfrak{n}M)$. Comparing the ranks as free S-modules, we consequently have $M \simeq S \otimes_k (M/\mathfrak{n}M)$. ∎

(10.2) REMARK. Note that F is never an equivalence of categories. Actually Hom-sets in $\mathfrak{M}(kG)$ are k-vector spaces, but not so in $\mathfrak{M}(S*G)$. However note that we obtained a very useful correspondence from (10.1):

There is a one-to-one correspondence between the set of irreducible representations of G over k and the set of indecomposable projective $S*G$-modules.

(10.3) DEFINITION. As before let V be a vector space of dimension two and let G be a finite subgroup of $\mathrm{GL}(2,k)$ acting on V in natural way. Suppose $\{V_0, V_1, \ldots, V_d\}$ is the set of all classes of non-isomorphic irreducible representations of G. For any representation W of G over k, we denote by $\mathrm{mult}_i(W)$ the dimension of $\mathrm{Hom}_{kG}(V_i, W)$ as a k-vector space. The **McKay graph** $\mathrm{Mc}(V, G)$ is defined to be an oriented graph whose vertices are V_i ($0 \le i \le d$) and there are μ arrows from V_i to V_j when $\mu = \mathrm{mult}_i(V \otimes_k V_j) \ne 0$.

Let $\{P_0, P_1, \ldots, P_d\}$ be the set of classes of indecomposable projective $S*G$-modules such that $F(V_i) = P_i$ ($0 \le i \le d$) with the notation in (10.1). For any projective $S*G$-module P we define an integer $\nu_i(P)$ ($0 \le i \le d$) to be the number of copies of P_i appearing in the direct decomposition of P. We can prove:

(10.4) LEMMA. $\mathrm{mult}_i(V \otimes_k V_j) = \nu_i(F(V \otimes_k V_j))$ ($0 \le i, j \le d$).

PROOF: By definition $V \otimes_k V_j = \sum_i V_i^{(\mu_i)}$ where $\mu_i = \mathrm{mult}_i(V \otimes_k V_j)$. Therefore it follows that $F(V \otimes_k V_j) = \sum_i P_i^{(\mu_i)}$. ∎

Keeping the above notation, let us now study the subring $R = S^G$ of invariants. As in the previous chapter denote by $\mathfrak{C}(R)$ the category of CM modules over R. Recall that an R-module is an object in $\mathfrak{C}(R)$ only when it is reflexive, because R is a normal domain of dimension two, (1.5.3). In our case we have a very useful result to study the category.

(10.5) PROPOSITION. *Let $\mathrm{add}_R(S)$ be the full subcategory of $\mathfrak{M}(R)$ whose objects are isomorphic to R-summands of free S-modules. Then, as a subcategory of $\mathfrak{M}(R)$, $\mathrm{add}_R(S)$ is equal to $\mathfrak{C}(R)$. In particular, we can identify the set of indecomposable CM modules* over R and the set of indecomposable R-summands of S. Consequently R is of finite representation type.

PROOF: Since S is reflexive as an R-module, it is clear that $\mathrm{add}_R(S) \subset \mathfrak{C}(R)$. To see the equality let M be a CM module over R. Note that the natural embedding of R into S is a split monomorphism as an R-module, since the R-homomorphism $\varphi : S \to R$ defined by $\varphi(s) = \frac{1}{|G|}\sum\{\sigma(s)|\sigma \in G\}$ gives a retraction. Applying $\mathrm{Hom}_R(M^*,\ )$ to this, we have a split monomorphism $\mathrm{Hom}_R(M^*, R) \to \mathrm{Hom}_R(M^*, S)$, where $M^* = \mathrm{Hom}_R(M, R)$. Since M is a reflexive R-module, we have $\mathrm{Hom}_R(M^*, R) \simeq M$. On the other hand, since $\mathrm{Hom}_R(M^*, S)$ is reflexive as an S-module and since S is regular, we see that $\mathrm{Hom}_R(M^*, S)$ is a free S-module. We thus conclude that M is an R-summand of a free S-module, hence $\mathrm{add}_R(S) = \mathfrak{C}(R)$. Moreover if M is indecomposable, then M must be a summand of the R-module S by Krull-Schmidt theorem, and the second statement of the lemma follows. ∎

Next we investigate the relation between $\mathfrak{C}(R)$ and $\mathfrak{M}(S * G)$. Before this, we make several remarks on pseudo-reflections in $\mathrm{GL}(2, k)$.

(10.6) DEFINITION. An element σ in $\mathrm{GL}(2, k)$ is said to be a **pseudo-reflection** if

$$\mathrm{rank}(\sigma - 1) \leq 1.$$

As to pseudo-reflections the following facts are basic.

(10.7) LEMMA. *Let G be a finite subgroup of $\mathrm{GL}(2, k)$. Suppose that G acts naturally on $S = k\{x, y\}$ and that the invariant subring is R. Then,*

(10.7.1) *R is regular if and only if G is generated by pseudo-reflections.*

(10.7.2) *The ring extension $R \subset S$ is unramified in codimension one if and only if G has no pseudo-reflection except the identity.* (By definition, $R \subset S$ is unramified in codimension one if, for any prime ideal $\mathfrak{P}$ of S with height ≤ 1, $(\mathfrak{P} \cap R)S_{\mathfrak{P}} = \mathfrak{P} S_{\mathfrak{P}}$.)

PROOF: We omit the proof of (10.7.1); see Bourbaki [**17**, §5, no. 5] for example.

To prove (10.7.2) let

$$T_{\mathfrak{P}} = \{\sigma \in G|\ \sigma(a) - a \in \mathfrak{P} \text{ for any } a \in S\}$$

for a prime $\mathfrak{P}$ of S with height ≤ 1, and call it the inertia group at $\mathfrak{P}$. By Serre [**59**, I, Proposition 20] it is known that

$$\sharp T_{\mathfrak{P}} = \mathrm{length}_{S_{\mathfrak{P}}}(S_{\mathfrak{P}}/(\mathfrak{P} \cap R)S_{\mathfrak{P}}).$$

Therefore S is unramified in codimension one over R only when $T_{\mathfrak{P}} = \{1\}$ for any $\mathfrak{P}$ of height one. It is, thus, enough to show that $\sigma \in G$ is a pseudo-reflection if and only if

$\sigma \in T_{\mathfrak{P}}$ for some $\mathfrak{P}$ of height one. If $\sigma \in G$ is a pseudo-reflection then V has a basis $\{x, y\}$ with $\sigma(x) = \zeta x$, $\sigma(y) = y$ ($\zeta \in k$). Then σ acts trivially on S/xS, and hence $\sigma \in T_{\mathfrak{P}}$ with $\mathfrak{P} = xS$. Conversely, assume $\sigma \in T_{\mathfrak{P}}$ for some $\mathfrak{P}$ of height one. Since S is a unique factorization domain, we can find $z \in S$ with $\mathfrak{P} = zS$. If $z \in \mathfrak{n}^2$, where $\mathfrak{n}$ is the maximal ideal of S, then σ acts trivially on $S/\mathfrak{n}^2$ and hence does so on V. Thus $\sigma = 1$. If $z \notin \mathfrak{n}^2$, then, since the action of $\sigma - 1$ on $\mathfrak{n}/zS + \mathfrak{n}^2$ is null, we see that the rank of $(\sigma - 1)$ on V is not more than one, and hence σ is a pseudo-reflection. ∎

Furthermore we have the following lemma.

(10.8) LEMMA. *Keeping the above notation, let* $\delta : S * G \to \mathrm{End}_R(S)$ *be an R-algebra mapping defined by*
$$\delta(s\sigma)(t) = s\sigma(t) \quad (\sigma \in G \text{ and } s, t \in S)$$
If the extension $R \subset S$ is unramified in codimension one, then δ is an isomorphism.

PROOF: (i) Firstly we show that δ is an isomorphism in the case when $R = S^G \subset S$ is a separable extension.

By definition the extension $R \subset S$ is separable if there is an element $e = \sum_i x_i \otimes y_i$ in $S \otimes_R S$ so that

(*) $$\sum_i x_i y_i = 1 \quad \text{and} \quad (1 \otimes a - a \otimes 1)e = 0 \quad \text{for any } a \in S.$$

Note that the assumption of the lemma means that the extension $R_{\mathfrak{p}} \subset S_{\mathfrak{p}}$ is separable for each prime $\mathfrak{p}$ of R of height ≤ 1. In the separable case it can be seen that

(**) $$\sum_i x_i \sigma(y_i) = \begin{cases} 1 & (\sigma = 1 \in G), \\ 0 & (\sigma \neq 1 \in G). \end{cases}$$

In fact, from (*) we have $\sum_i x_i a \otimes y_i = \sum_i x_i \otimes a y_i$ for any $a \in S$. Applying the automorphism $1 \otimes \sigma$ to this, we obtain $\sum_i x_i a \otimes \sigma(y_i) = \sum_i x_i \otimes \sigma(a)\sigma(y_i)$. Letting $e_\sigma = \sum_i x_i \sigma(y_i)$ we thus have $a e_\sigma = \sigma(a) e_\sigma$ for any $a \in S$. Hence $e_\sigma = 0$ if $\sigma \neq 1$.

Now let h be an arbitrary element in $\mathrm{End}_R(S)$. Setting $z = \sum_i \sum_\sigma h(x_i)\sigma(y_i)\sigma \in S * G$, we obtain the equalities for any $a \in S$:

$$(\delta z)(a) = \sum_{\sigma, i} h(x_i)\sigma(y_i)\sigma(a) = \sum_i h(x_i)(\sum_\sigma \sigma(ay_i))$$
$$= h(\sum_i x_i(\sum_\sigma \sigma(ay_i))) = h(\sum_\sigma (\sum_i x_i \sigma(y_i))\sigma(a))$$
$$= h(a).$$

(We used (**) for the last equality.) We have thus shown $\delta(z) = h$. Therefore it follows that δ is an epimorphism. Comparing the ranks of $\text{End}_R(S)$ and $S * G$ as R-modules, we conclude that δ must be an isomorphism in this particular case.

(ii) Now we prove the lemma. Let K, L be the field of quotients of R, S respectively. Note that $K \subset L$ is a Galois extension with Galois group G, in particular, it is separable. Hence by (i) we know that the natural mapping $\delta \otimes_R K : L * G \to \text{End}_K(L)$ is an isomorphism. Since we have the commutative diagram

$$\begin{array}{ccc} S * G & \xrightarrow{\delta} & \text{End}_R(S) \\ \cap & & \cap \\ L * G & \xrightarrow{\delta \otimes_R K} & \text{End}_K(L), \end{array}$$

it follows that δ is a monomorphism. To see the lemma, it is thus enough to show that $\delta \otimes_R R_{\mathfrak{p}}$ is an isomorphism for each prime $\mathfrak{p}$ in R of height one, since both $S * G$ and $\text{End}_R(S)$ are reflexive R-modules. However this is just what we have shown in (i), because $R_{\mathfrak{p}} \subset S_{\mathfrak{p}}$ is a separable extension. ∎

We now have a striking relation between CM modules over R and $S * G$-modules.

(10.9) PROPOSITION. (Auslander [6]) *Keeping the above notation, assume G has no pseudo-reflection but the identity. For an $S*G$-module M and for an $S*G$-homomorphism $f : M \to N$, let $H(M) = M^G$ and $H(f) = f|_{M^G}$. Then the functor H yields the equivalence of categories:*

$$\wp(S * G) \simeq \mathfrak{C}(R).$$

PROOF: (i) First we have to verify that M^G is actually a CM module over R when M is a projective $S * G$-module.

It is clear that M^G is a direct summand of M as an R-module. Since M is projective over $S * G$, it is also a projective S-module, hence it is S-free. Therefore M^G belongs to $\text{add}_R(S)$ and our contention follows from (10.5).

(ii) Note that G acts on $S * G$ as follows:

$$\sigma(\sum_i a_i \tau_i) = \sum_i \sigma(a_i) \sigma \tau_i \quad (a_i \in S, \sigma, \tau_i \in G).$$

Then it is easy to see that the invariant part under this action is

$$(S * G)^G = \{\sum_\sigma \sigma(s)\sigma | s \in S\}.$$

We denote the R-submodule $(S * G)^G$ of $S * G$ by S_1. Note that $S_1 \simeq S$ as an R-module.

(iii) We show that a restriction mapping $\alpha : \mathrm{End}_{S*G}(S*G) \to \mathrm{End}_R(S_1)$ defined by $\alpha(f) = f|_{S_1}$ ($f \in \mathrm{End}_{S*G}(S*G)$) gives an isomorphism of R-algebras.

To see this, define a sequence of algebra mappings:

$$S*G \xrightarrow{\gamma} (S*G)^{op} \xrightarrow{\beta} \mathrm{End}_{S*G}(S*G) \xrightarrow{\alpha} \mathrm{End}_R(S_1),$$

by equalities:

$$\beta(\xi)(\eta) = \eta \cdot \xi \qquad (\xi, \eta \in S*G),$$
$$\gamma(s\sigma) = \sigma^{-1}(s)\sigma^{-1} \qquad (s \in S, \sigma \in G).$$

It is an easy exercise to see that β and γ are bijective and that the composition $\alpha \cdot \beta \cdot \gamma$ is equal to δ defined in (10.8), where we identify S_1 with S as in (ii). It then follows from (10.7.2) and from (10.8) that δ, hence α, is an isomorphism.

(iv) In general, let $A \subset B$ be an extension of (noncommutative) rings where B is finite as a left A-module. We denote by $\mathrm{add}_A(B)$ the category of left A-modules which are A-summands of free B-modules. Then each object P in $\mathrm{add}_A(B)$ has an exact sequence of A-modules:

$$B^{(\nu)} \xrightarrow{f=(f_{ij})} B^{(\nu)} \xrightarrow{h} P \longrightarrow 0,$$

where $f_{ij} \in \mathrm{End}_A(B)$ and the restriction of f onto $\mathrm{Ker}(h) = \mathrm{Im}(f)$ is the identity mapping. Define a category $P(\mathrm{End}_A(B))$, whose objects are square matrices f on $\mathrm{End}_A(B)$ with the property $f^2 = f$, and the morphisms between f and f' are pairs (α, β) of A-homomorphisms with the commutative diagram

$$\begin{array}{ccc} B^{(\nu)} & \xrightarrow{f} & B^{(\nu)} \\ \beta \downarrow & & \alpha \downarrow \\ B^{(\nu')} & \xrightarrow{f'} & B^{(\nu')}, \end{array}$$

modulo the set $\{(f'\cdot\gamma, \gamma\cdot f) | \gamma \in \mathrm{Hom}_A(B^{(\nu)}, B^{(\nu')})\}$. Under these circumstances it is easy to see that the category $\mathrm{add}_A(B)$ is equivalent to $P(\mathrm{End}_A(B))$. (Exercise: Prove this. Note that the equivalence is given by $Q : P(\mathrm{End}_A(B)) \to \mathrm{add}_A(B); Q(f) = \mathrm{Coker}(f)$.)

(v) With the notation in (iv) we note that $\mathrm{add}_{S*G}(S*G) = \wp(S*G)$. On the other hand we know from (10.5) that $\mathfrak{C}(R) = \mathrm{add}_R(S)$. Hence we have

$$\wp(S*G) \simeq P(\mathrm{End}_{S*G}(S*G)),$$
$$\mathfrak{C}(R) \simeq P(\mathrm{End}_R(S)).$$

Since $\mathrm{End}_{S*G}(S*G) \simeq \mathrm{End}_R(S)$ as an R-algebra by (iii), we obtain the equivalence of categories $\mathfrak{C}(R) \simeq \wp(S*G)$ as required. We leave the reader to check that the equivalence is actually given by the functor H in the lemma. ∎

Combining this with (10.2) we have the following

(10.10) COROLLARY. *The composition $H \cdot F$ of functors yields a one-to-one correspondence between the set of classes of irreducible representations of G and the set of classes of indecomposable CM modules over R.*

Making use of this correspondence we will show that the AR quiver is the same as the McKay graph. We arrange the notation first.

(10.11) NOTATION. In the rest of this chapter $G \subset \mathrm{GL}(2,k)$ is supposed to have no pseudo-reflections but the identity. Let $\{V_0, V_1, \ldots, V_d\}$ be the set of nonisomorphic simple kG-modules, where V_0 is the trivial kG-module k. Also we denote by $\{P_0, P_1, \ldots, P_d\}$ the set of indecomposable projective $S*G$-modules with $P_i = F(V_i)$ ($0 \leq i \leq d$). We define the permutations on these sets as follows:

$$\tau(V_i) = \bigwedge^2 V \otimes_k V_i$$
$$\tau(P_i) = F(\tau(V_i)) \quad (0 \leq i \leq d).$$

Note that $\tau(V_i) \simeq \tau(V_j) \iff \tau(P_i) \simeq \tau(P_j) \iff i = j$.

Let $\{L_0, L_1, \ldots, L_d\}$ be the set of CM modules over R where $L_i = H(P_i)$ with the notation in (10.9). Note that $L_0 = R$. We also denote by τ the permutation on this set induced from τ on the P_i. In other words,

$$\tau(L_i) = H(\tau(P_i)) \quad (0 \leq i \leq d).$$

(10.12) REMARK. With the above notation, $\tau(L_0)$ is isomorphic to the canonical module of R. In fact $\tau(L_0)$ can be described as $(S \otimes_k \wedge^2 V)^G$ where the action of G on $(S \otimes_k \wedge^2 V)$ is given by $\sigma(s \otimes (v \wedge w)) = \sigma(s) \otimes det(\sigma)(v \wedge w)$ ($\sigma \in G, s \in S$ and $v, w \in V$). Therefore $\tau(L_0) = \{s \in S|\ \sigma(s) = (1/det(\sigma))s$ for any $\sigma \in G\}$, and this is isomorphic to the canonical module by Watanabe [64].

Writing the Koszul complex over S as

(10.13.1) $$0 \to S \otimes_k \bigwedge^2 V \to S \otimes_k V \to S \to k \to 0,$$

we easily see that this is also an exact sequence of $S*G$-modules. Applying the functor $\otimes_k V_i$ to this, we obtain

(10.13.2) $\quad 0 \to S \otimes_k (\bigwedge^2 V \otimes_k V_i) \to S \otimes_k (V \otimes_k V_i) \to S \otimes_k V_i \to V_i \to 0,$

which gives the minimal projective resolution of the $S*G$-module V_i. Using the notation of (10.11), we hence have an exact sequence:

$$0 \to \tau(P_i) \to F(V \otimes_k V_i) \to P_i \to V_i \to 0 \quad (0 \le i \le d).$$

Now take the G-invariants of this sequence to get

(10.13.3) $\quad 0 \to \tau(L_i) \to H \cdot F(V \otimes_k V_i) \to L_i \to V_i^G \to 0.$

Note here that $V_i^G = k$ if $i = 0$ and $V_i^G = 0$ otherwise, because each V_i is a simple kG-module. Consequently the sequences (10.13.3) become

(10.13.4) $\quad \begin{array}{ll} 0 \to \tau(L_0) \to E_0 \xrightarrow{p_0} L_0 \to k \to 0 & (i=0), \\ 0 \to \tau(L_i) \to E_i \xrightarrow{p_i} L_i \to 0 & (i \ne 0), \end{array}$

where $E_i = H \cdot F(V \otimes_k V_i)$.

(10.13) PROPOSITION. *For any i $(0 \le i \le d)$, the sequences (10.13.4) satisfy the following condition:*

If L is a CM module over R and if $f : L \to L_i$ is an R-homomorphism which is not a split epimorphism, then there exists an R-homomorphism $g : L \to E_i$ with $f = p_i \cdot g$.

In particular, if $i \ne 0$ then the sequence (10.13.4) is the AR sequence ending in L_i.

PROOF: Recall the equivalence between $\wp(S*G)$ and $\mathfrak{C}(R)$ is given by the functor H in (10.9). For a given $f : L \to L_i$ consider the corresponding diagram in $\wp(S*G)$:

$$\begin{array}{ccc} F(V \otimes_k V_i) & \xrightarrow{H^{-1}(p_i)} & P_i \\ & & \uparrow H^{-1}(f) \\ & & H^{-1}(L) \end{array}$$

Since f is not a split epimorphism, we have $\text{Im}(f) \subseteq \text{Im}(p_i)$, hence it follows that $\text{Im}(H^{-1}(f)) \subseteq \text{Im}(H^{-1}(p_i))$. By this together with the fact that $H^{-1}(L)$ is a projective

$S * G$-module, we show there is an $S * G$-homomorphism g' from $H^{-1}(L)$ into $F(V \otimes_k V_i)$ so that $H^{-1}(f) = H^{-1}(p_i) \cdot g'$. Letting $g = H(g')$ we have $f = p_i \cdot g$ as required. ∎

Now recall the definition of the AR quiver Γ for R. Vertices in Γ are the indecomposable CM modules over R and hence they are L_i ($0 \leq i \leq d$). By the above lemma, if $i \neq 0$, then $\mathrm{irr}(L_j, L_i)$ is equal to the number of copies of L_j appearing in the direct decomposition of E_i; see (5.5). Even in the case when $i = 0$, the sequence (10.13.4) has the same property as AR sequences, therefore we can verify the same equality as above. By using the equivalence between $\wp(S * G)$ and $\mathfrak{C}(R)$, we thus have shown that $\mathrm{irr}(L_j, L_i) = \nu_j(F(V \otimes_k V_i))$, which is equal to $\mathrm{mult}_j(V \otimes_k V_i)$ by (10.4). This shows that the map $\Gamma \to \mathrm{Mc}(V, G)$ sending $[L_i]$ to $[V_i]$ yields an isomorphism of graphs. We therefore have proved:

(10.14) **THEOREM.** (Auslander [6]) *If $G \subset \mathrm{GL}(2, k)$ has no pseudo-reflection but the identity, then the invariant subring $R = S^G$ is always of finite representation type, and the AR quiver Γ of R coincides with the McKay graph $\mathrm{Mc}(V, G)$.*

We end this chapter by giving several examples of AR quivers.

(10.15) *Important Examples.*

Let ζ_n be the primitive n-th root of unity in an algebraically closed field k of characteristic 0. It is a classical result that any finite subgroup of $\mathrm{SL}(2, k)$ is conjugate to one of the following groups (called **Klein groups**):

(A_n) Cyclic group of order $n + 1$;

$$C_n = < \begin{pmatrix} \zeta_{n+1} & 0 \\ 0 & \zeta_{n+1}^{-1} \end{pmatrix} >.$$

(D_n) Binary dihedral group of order $4(n-2)$;

$$D_n = < \begin{pmatrix} 0 & \zeta_4 \\ \zeta_4 & 0 \end{pmatrix}, \ C_{2n-5} >.$$

(E_6) Binary tetrahedral group of order 24;

$$T = < \frac{1}{\sqrt{2}} \begin{pmatrix} \zeta_8 & \zeta_8^3 \\ \zeta_8 & \zeta_8^7 \end{pmatrix}, \ D_4 >.$$

(E_7) Binary octahedral group of order 48;

$$O = < \begin{pmatrix} \zeta_8^3 & 0 \\ 0 & \zeta_8^5 \end{pmatrix}, \ T >.$$

(E_8) Binary icosahedral group of order 120;

$$I = <\frac{1}{\sqrt{5}}\begin{pmatrix} \zeta_5^4 - \zeta_5 & \zeta_5^2 - \zeta_5^3 \\ \zeta_5^2 - \zeta_5^3 & \zeta_5 - \zeta_5^4 \end{pmatrix}, \frac{1}{\sqrt{5}}\begin{pmatrix} \zeta_5^2 - \zeta_5^4 & \zeta_5^4 - 1 \\ 1 - \zeta_5 & \zeta_5^3 - \zeta_5 \end{pmatrix}>.$$

It was also proved by Klein that the invariant subrings by these groups are simple hypersurface singularities, i.e., $S^G \simeq k\{x, y, z\}/(f)$ where f is one of the following polynomials respectively in each case:

(A_n) $x^2 + y^{n+1} + z^2$ $(n \geq 1)$,

(D_n) $x^2 y + y^{n-1} + z^2$ $(n \geq 4)$,

(E_6) $x^3 + y^4 + z^2$,

(E_7) $x^3 + xy^3 + z^2$,

(E_8) $x^3 + y^5 + z^2$.

The AR quivers for these rings are computed as McKay graphs of the above group. We exhibit them now. Each number attached to a vertex indicates the rank of the corresponding CM module. When we regard them as McKay graphs, the numbers are the degrees of irreducible representations.

(A_n)

$$[R]$$
$$1 \rightleftarrows 1 \rightleftarrows 1 \rightleftarrows \ldots \rightleftarrows 1 \rightleftarrows 1 \rightleftarrows 1$$

(D_n)

$$[R]$$
$$2 \rightleftarrows 2 \rightleftarrows 2 \rightleftarrows \ldots \rightleftarrows 2 \rightleftarrows 2 \begin{matrix} 1 \\ 1 \end{matrix}$$
$$1$$

(E_6)

$$[R]$$
$$\updownarrow$$
$$2$$
$$\updownarrow$$
$$1 \rightleftarrows 2 \rightleftarrows 3 \rightleftarrows 2 \rightleftarrows 1$$

(E_7)

$$2$$
$$\updownarrow$$
$$[R] \rightleftarrows 2 \rightleftarrows 3 \rightleftarrows 4 \rightleftarrows 3 \rightleftarrows 2 \rightleftarrows 1$$

(E_8)
$$[R] \rightleftarrows 2 \rightleftarrows 3 \rightleftarrows 4 \rightleftarrows 5 \rightleftarrows 6 \rightleftarrows 4 \rightleftarrows 2$$
with a $\Updownarrow 3$ above the 6.

(10.16) *Examples of invariant subrings by cyclic groups.*

Let G be a cyclic subgroup of $GL(2, k)$. Then the McKay graph $Mc(V, G)$ is easily computed, therefore the AR quiver will be obtained.

Let R_1 be a Veronese subring $k\{x^n, x^{n-1}y, x^{n-2}y^2, \ldots, y^n\}$ of degree n which is an invariant part of $k\{x, y\}$ by action of a cyclic group of order n. The AR quiver for R_1 is shown in Figure (10.16.1), where K denotes the class of the canonical module of R.

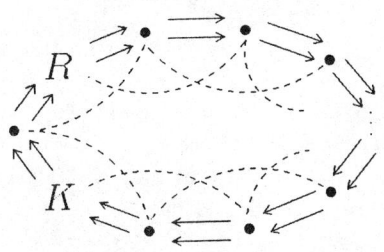

Figure (10.16.1)

Figures (10.16.2) and (10.16.3) show the AR quivers for the ring $R_2 = k\{x^5, x^3y, xy^2, y^5\}$ and $R_3 = k\{x^{10}, x^7y, x^4y^2, xy^3, y^{10}\}$ respectively.

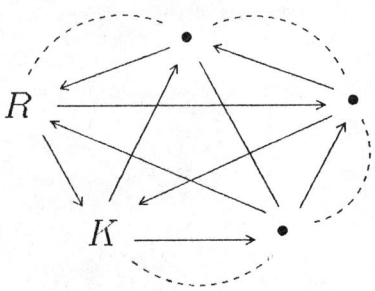

Figure (10.16.2)

McKay graphs

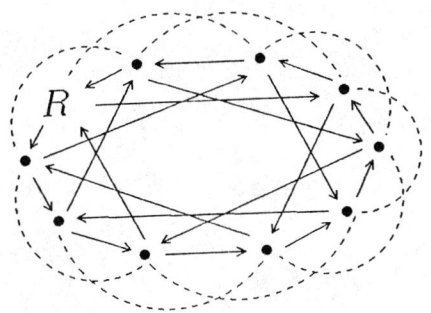

Figure (10.16.3)

Chapter 11. Two-dimensional CM rings of finite representation type

In the last chapter we showed that the AR quiver of an invariant subring of dimension two is the same thing as the McKay graph of a finite group. One of the main purposes here is to show that any CM ring of dimension two of finite representation type is a ring of invariants by a finite group; see Theorem (11.4). Therefore all the finite AR quivers for CM rings of dimension two are obtained as McKay graphs.

Throughout this chapter $(R, \mathfrak{m}, k)$ is an analytic *normal* local domain of dimension two (so that it is CM) with k an *algebraically closed field of characteristic* 0 (or $k = \mathbb{C}$), and we denote by Q the field of quotients of R. Let L be a finite Galois extension of Q with Galois group G and let S be the integral closure of R in L. Then S is a local ring which is module-finite over R. Denoting by $\mathfrak{n}$ the maximal ideal of S, we note $S/\mathfrak{n} \simeq k$, since k is algebraically closed. As before $\mathfrak{C}(R)$ denotes the category of CM modules over R and $\mathrm{add}_R(S)$ is the category consisting of R-summands of free S-modules. Note that all modules belonging to $\mathfrak{C}(R)$ are reflexive and conversely any reflexive modules are in $\mathfrak{C}(R)$. Furthermore one can easily see that $\mathrm{add}_R(S)$ is a (full) subcategory of $\mathfrak{C}(R)$. Also denote by $\wp(S * G)$ (resp. $\mathfrak{M}(kG)$) the category of projective left $S*G$-modules (resp. left kG-modules). The proofs of Lemma (10.1) and (10.9) can be easily imitated to yield the following two lemmas. (The difference here is that S might not be regular. However we did not use this in the proof of (10.1) and (10.9).)

(11.1) LEMMA. *The functor* $F : \mathfrak{M}(kG) \to \wp(S*G)$ *defined by* $F(W) = W \otimes_k S$ ($W \in \mathfrak{M}(kG)$) *gives a one-to-one correspondence between the sets of classes of objects in these two categories.*

(11.2) LEMMA. *Suppose that the ring extension* $R \subset S$ *is unramified in codimension one. Then the functor* $H : \wp(S*G) \to \mathrm{add}_R(S)$; $H(M) = M^G$ ($M \in \wp(S*G)$) *is an equivalence of categories.*

Exercise. Give complete proofs of these lemmas.

The following lemma due to Mumford will be necessary in the proof of our theorem.

(11.3) LEMMA. *Keeping the notation as above, we assume R satisfies the following condition:*

(11.3.1) *For any finite Galois extension L of Q, if the integral closure S of R in L is unramified in codimension one over R, then $S = R$. That is, R has no nontrivial extensions unramified in codimension one.*
Then R is regular, i.e. $R \simeq k\{x,y\}$ for some x and y.

For the proof we refer the reader to Mumford [49].

We are now able to prove the theorem. We should mention that the following algebraic proof is due to Auslander [6], and there is another, more geometric, proof given by Esnault [27].

(11.4) THEOREM. *With the above notation, suppose R is of finite representation type. Then there exists a finite Galois extension Ω of Q with Galois group G so that the integral closure S of R in Ω is isomorphic to $k\{x,y\}$ on which G acts linearly, and $R = S^G$.*

PROOF: Consider the following set of field extensions of Q in some fixed algebraic closure of Q:

$$\Delta = \{L|\ L \text{ is a finite extension of the field } Q \text{ such that the integral closure } S_L \text{ of } R \text{ in } L \text{ is unramified in codimension one over } R\}.$$

Setting $\Omega = \cup\{L \in \Delta\}$, the smallest field containing all L in Δ, we can easily check that Ω is a Galois extension of Q. We will show that

(11.4.1) Ω is a finite extension of Q.

For, if $[\Omega : Q] = \infty$, then there is a series of finite Galois extensions of Q:

$$Q \subset L_1 \subset L_2 \subset \ldots \subset L_n \subset L_{n+1} \subset \ldots \subset \Omega.$$

Let G_n be the Galois group of the extension $Q \subset L_n$ and let S_n be the integral closure of R in L_n. Since S_n is a direct summand of S_{n+1} as an S_n-module, we have a sequence of full subcategories of $\mathfrak{C}$:

$$\text{add}_R(S_1) \subset \text{add}_R(S_2) \subset \ldots \subset \text{add}_R(S_n) \subset \ldots \subset \mathfrak{C}(R)$$

The assumption that $\mathfrak{C}(R)$ has only a finite number of indecomposable objects implies that, for n large enough, $\text{add}_R(S_n)$ and $\text{add}_R(S_{n+1})$ have the same number of indecomposable modules. It then turns out from (11.1) and (11.2) that the number of simple

kG_n-modules is equal to that of simple kG_{n+1}-modules. In particular G_n and G_{n+1} have the same number of conjugacy classes. On the other hand, since L_n is an intermediate field of $Q \subset L_{n+1}$, we have a normal subgroup H of G_{n+1} with $G_n \simeq G_{n+1}/H$. Comparing the number of conjugacy classes of G_n and G_{n+1}, we conclude that $G_n = G_{n+1}$, hence $L_n = L_{n+1}$. This contradicts the choice of L_n, and (11.4.1) follows.

Now let S be the integral closure of R in Ω and G the Galois group of $Q \subset \Omega$. By (11.4.1) G is a finite group and $R = S^G$. Note that by definition S has no ring extension unramified in codimension one over S. It thus follows from (11.3) that $S = k\{x, y\}$ for some x and y. It is well-known, and easy to see, that the action of G on S can be linearized, that is, after changing variables, G may act on the vector space $kx \oplus ky$. This completes the proof of the theorem. ∎

By this theorem we see that two-dimensional CM rings of finite representation type are the ones discussed in the previous chapter.

Next we shall briefly explain how to construct the AR quivers for general normal domains of dimension two.

(11.5) DEFINITION. Let $(R, \mathfrak{m}, k)$ be an analytic normal local domain of dimension two. By (1.11) there exists a canonical module K_R over R. By definition K_R is a module with the property $\operatorname{Ext}_R^2(k, K_R) \simeq k$, or equivalently $\operatorname{Ext}_R^1(\mathfrak{m}, K_R) \simeq k$. This means that there is a unique nonsplit exact sequence:

(11.5.1) $$0 \longrightarrow K_R \longrightarrow E \longrightarrow \mathfrak{m} \longrightarrow 0.$$

We call this the **fundamental sequence** of R. Here the module E appearing in the middle term is also unique up to isomorphism, which is called the **fundamental module** of R.

(11.6) LEMMA. *The fundamental module E is a reflexive module of rank 2, which is generated by at most $\operatorname{r}(R) + \operatorname{emb}(R)$ elements, where $\operatorname{r}(R)$ is the Cohen-Macaulay type of R and $\operatorname{emb}(R)$ denotes the embedding dimension of R. Furthermore there is an isomorphism of R-modules:*

$$(\bigwedge^2 E)^{**} \simeq K_R.$$

PROOF: Applying the functor $\operatorname{Hom}_R(\ , K_R)$ to the fundamental sequence, we get the exact sequence:

$$\begin{aligned}&\to \operatorname{Hom}_R(K_R, K_R) \xrightarrow{\alpha} \operatorname{Ext}_R^1(\mathfrak{m}, K_R) \to \operatorname{Ext}_R^1(E, K_R) \to \operatorname{Ext}_R^1(K_R, K_R) \\ &\to \operatorname{Ext}_R^2(\mathfrak{m}, K_R) \to \operatorname{Ext}_R^2(E, K_R) \to \operatorname{Ext}_R^2(K_R, K_R) \to \cdots\end{aligned}$$

where α is surjective by definition and $\text{Ext}^1_R(K_R, K_R) = \text{Ext}^2_R(\mathfrak{m}, K_R) = \text{Ext}^2_R(K_R, K_R) = 0$ by the local duality theorem. Hence we obtain $\text{Ext}^1_R(E, K_R) = \text{Ext}^2_R(E, K_R) = 0$ and this implies that depth$(E) = 2$ and that E lies in $\mathfrak{C}(R)$.

Taking the divisor classes in the sense of Bourbaki (or the first Chern classes) attached to the modules in the fundamental sequence, we obtain the equality $c(E) = c(K_R)$ in the divisor class group of R. (See Bourbaki [16,§7].) Then by definition it follows that $c(\Lambda^2 E) = c(K_R)$; cf. [16, Exercise 12 of §4]. This gives the isomorphism in the lemma.

By noting that $r(R) = \dim_k(K_R \otimes_R k)$ and emb$(R) = \dim_k(\mathfrak{m} \otimes_R k)$, the remaining part of the lemma follows from the fundamental sequence (11.5.1). ∎

(11.7) COROLLARY. *Assume that R is Gorenstein. Then R is a regular local ring if and only if the fundamental module E has a free direct summand.*

PROOF: If R is regular, then the fundamental sequence is given by

$$0 \longrightarrow R \longrightarrow R^2 \longrightarrow \mathfrak{m} \longrightarrow 0.$$

In particular, $E \simeq R^2$. Conversely assume that $E \simeq R \oplus \mathfrak{a}$ for some ideal $\mathfrak{a}$. Since R is Gorenstein so that $K_R \simeq R$, it then follows from (11.6) that $R \simeq (\Lambda^2 E)^{**} \simeq \mathfrak{a}^{**} \simeq \mathfrak{a}$, hence $E \simeq R^2$. This together with the exact sequence $0 \to R \to E \to \mathfrak{m} \to 0$ shows that R is regular. ∎

(11.8) *Example.* Let V be a k-vector space of dimension two with basis $\{x, y\}$ and let G be a finite subgroup of GL(V). Then we have a natural action of G on the regular local ring $S = k\{x, y\}$. The fundamental module of the invariant subring $R = S^G$ will be obtained as follows: Define the action of G on the free module $S \otimes_k V$ of rank 2 by

$$\sigma(s \otimes v) = \sigma(s) \otimes \sigma(v) \quad (\sigma \in G, s \in S, v \in V).$$

Denote by E the invariant part of this action:

$$E = (S \otimes_k V)^G.$$

Then E is the fundamental module of R. Indeed, we know from (10.12) and (10.13) that the sequence (10.13.4), when $i = 0$, is the fundamental sequence of R.

(11.9) *Exercise.* Prove that if R is a hypersurface, then the fundamental module E is isomorphic to the third syzygy of k. See Yoshino-Kawamoto [68] for more details.

Considering $\mathfrak{m} \subset R$ in the fundamental sequence we rewrite (11.5.1) as follows:

$$\sigma : 0 \longrightarrow K_R \xrightarrow{q} E \xrightarrow{p} R.$$

Then the sequence has the property of an AR sequence:

(11.10) LEMMA.
(11.10.1)   Let $f : M \to R$ be an R-homomorphism of a CM module M into R which is not a split epimorphism, then there is an R-homomorphism $g : M \to E$ with $f = p \cdot g$.
(11.10.2)   If M is a CM module over R that is not free, then the following is exact:

$$\sigma(M) : 0 \longrightarrow \operatorname{Hom}_R(M^*, K_R) \stackrel{\operatorname{Hom}(M^*,q)}{\longrightarrow} \operatorname{Hom}_R(M^*, E) \stackrel{\operatorname{Hom}(M^*,p)}{\longrightarrow} M \longrightarrow 0.$$

PROOF: (11.10.1): The mapping f yields $\operatorname{Ext}^1_R(f, K_R) : \operatorname{Ext}^1_R(\mathfrak{m}, K_R) \to \operatorname{Ext}^1_R(M, K_R)$. Denoting $\tau = \operatorname{Ext}^1_R(f, K_R)(\sigma)$, we have a commutative diagram

$$\begin{array}{ccccccccc}
\tau : 0 & \longrightarrow & K_R & \longrightarrow & M \times_R E & \longrightarrow & M & \longrightarrow & 0 \\
& & \| & & \downarrow & & \downarrow f & & \\
\sigma : 0 & \longrightarrow & K_R & \longrightarrow & E & \longrightarrow & R & &
\end{array}$$

with exact rows, where the right square is a pull-back diagram. Since $\operatorname{Ext}^1_R(M, K_R) = 0$ by the local duality theorem, it follows that $\tau = 0$, hence the sequence τ splits, which just means that there is $g : M \to E$ with $f = p \cdot g$.

(11.10.2): We may assume that M is indecomposable. Clearly it suffices to show that $\operatorname{Hom}_R(M^*, p)$ is an epimorphism. To do this, let f be an element of $\operatorname{Hom}_R(M^*, R) \simeq M$. Note that f is not a split epimorphism. For otherwise, M^* would contain R as a direct summand, thus $M \simeq R$, a contradiction. Therefore by (11.10.1) there is an element g in $\operatorname{Hom}_R(M^*, E)$ with $f = \operatorname{Hom}(M^*, p)(g)$, giving the required result. ∎

(11.11) THEOREM. (Auslander [6]) *If M is an indecomposable CM module over R which is not free, then the sequence $\sigma(M)$ in (11.10.2) is the AR sequence ending in M.*

PROOF: We divide the proof into several steps.

(i) Since M is a torsion free module, we can define the trace of an element f in $\operatorname{End}_R(M)$. More precisely, trace(f) is defined to be the trace of a linear mapping $f \otimes_R Q \in \operatorname{End}_Q(M \otimes_R Q)$ where Q is the field of quotients of R. It is easily seen that trace(f) is an element of Q that is integral over R, for M is finitely generated. Because R is integrally closed, we have trace$(f) \in R$. We thus obtain a mapping:

$$\operatorname{trace} : \operatorname{End}_R(M) \to R$$

Note that trace$(1_M) = \operatorname{rank}(M)$, in particular, the trace map is onto. Also note that *if* trace(f) *is a unit in R, then f is an automorphism on M.* To see this, let L be a sufficiently large finite field extension of Q so that L contains the all eigenvalues α_i ($1 \leq i \leq n$) of the linear mapping $f \otimes_R L$ and let v_i ($1 \leq i \leq n$) be eigenvectors with eigenvalue α_i. We may take each v_i from $M \otimes_R S$ where S is the integral closure of R in L. Note that the α_i are in S, since they are integral over R. Also note that S is a local ring. Let $\mathfrak{n}$ be the maximal ideal of S; then because $v_i \neq 0$, one can take a sufficiently large integer m such that each v_i is not contained in $\mathfrak{n}^m(M \otimes_R S)$. Suppose f is not an isomorphism. Then, since M is indecomposable, f is in the radical of $\operatorname{End}_R(M)$. Thus $f^r(M) \subset \mathfrak{m}M$ for some integer r, hence $(f \otimes_R S)^r(M \otimes_R S) \subset \mathfrak{n}(M \otimes_R S)$. It follows from this that $\alpha_i^{rm} v_i = (f \otimes S)^{rm}(v_i) \in \mathfrak{n}^m(M \otimes S)$, therefore each α_i must be a nonunit in S, and so is $\sum_i \alpha_i$. Hence we have $\operatorname{trace}(f) = \sum_i \alpha_i \in \mathfrak{n} \cap R = \mathfrak{m}$, a contradiction.

(ii) There is a natural isomorphism of functors on $\mathfrak{C}(R)$:

$$\operatorname{Hom}_R(\operatorname{Hom}_R(M,N),\ ) \simeq \operatorname{Hom}_R(N, \operatorname{Hom}_R(M^*,\ )).$$

In fact, it is known that there are natural homomorphims

$$\operatorname{Hom}_R(\operatorname{Hom}_R(M,N),\ ) \xrightarrow{\alpha} \operatorname{Hom}_R(N \otimes_R M^*,\ ) \xrightarrow{\beta} \operatorname{Hom}_R(N, \operatorname{Hom}_R(M^*,\ )),$$

where β is always an isomorphism. To see that α is an isomorphism, it is sufficient to show that the localized mapping $\alpha_\mathfrak{p}$ are isomorphisms for all primes $\mathfrak{p}$ of height one, since both modules are reflexive. However this is trivial since $\operatorname{Hom}_R(M,N)_\mathfrak{p} \simeq (N \otimes_R M^*)_\mathfrak{p}$.

(iii) We shall show that $\sigma(M)$ is a nonsplit sequence in the theorem. Suppose not. Then $\operatorname{Hom}_R(M, \operatorname{Hom}_R(M^*, p))$ would be an epimorphism, hence so would $\operatorname{Hom}_R(\operatorname{End}_R(M), p)$: $\operatorname{Hom}_R(\operatorname{End}_R(M), E) \to \operatorname{Hom}_R(\operatorname{End}_R(M), R)$ by (ii). However the trace mapping cannot be lifted to an element in $\operatorname{Hom}_R(\operatorname{End}_R(M), E)$, because trace $\in \operatorname{Hom}_R(\operatorname{End}_R(M), R)$ is a surjective mapping. This shows that $\sigma(M)$ is not split.

(iv) Suppose we are given a CM module X and a homomorphism $f : X \to M$ which is not a split epimorphism. We want to show that there is a homomorphism $g : X \to \operatorname{Hom}_R(M^*, E)$ such that $\operatorname{Hom}_R(M^*, p) \cdot g = f$. If this is true for any X and for any f, then $\sigma(M)$ is an AR sequence by (2.9). As in (ii) there is a commutative diagram:

$$\begin{array}{ccc} \operatorname{Hom}_R(X, \operatorname{Hom}_R(M^*, E)) & \xrightarrow{\operatorname{Hom}(X, \operatorname{Hom}(M^*, p))} & \operatorname{Hom}_R(X, \operatorname{Hom}_R(M^*, R)) \\ \uparrow & & \uparrow \gamma \\ \operatorname{Hom}_R(\operatorname{Hom}_R(M, X), E) & \xrightarrow{\operatorname{Hom}(\operatorname{Hom}(M, X), p)} & \operatorname{Hom}_R(\operatorname{Hom}_R(M, X), R) \end{array}$$

We denote by f' the element in $\operatorname{Hom}_R(\operatorname{Hom}_R(M, X), R)$ with $\gamma(f') = f$. We claim that f' is not a split epimorphism. For otherwise, there would be an element g in $\operatorname{Hom}_R(M, X)$

such that $f'(g) = 1$, which exactly says that $\text{trace}(f \cdot g) = 1$. Then by (i), $f \cdot g$ would be an automorphism on M. This contradicts f not being a split epimorphism, and so f' is not a split epimorphism. Then by (11.10.1) there is an element g' in $\text{Hom}_R(\text{Hom}_R(M, X), E)$ with $\text{Hom}_R(\text{Hom}_R(M, X), p)(g') = f'$. Denoting by g the element corresponding to g' in $\text{Hom}_R(X, \text{Hom}_R(M^*, E))$, we have $\text{Hom}_R(M^*, p) \cdot g = f$ as required. ∎

As an application of (11.11) we consider the indecomposability of fundamental modules. Let G, S and R be as in (11.8). The fundamental module E of R is given in (11.8). Furthermore assume G is a cyclic group. Then the action of G on V is diagonalized after a suitable choice of basis of V. So we have $E = (Sx)^G \oplus (Sy)^G$. Thus the fundamental module is decomposable in this case. We can prove the converse is true as well by using Theorem (11.11).

(11.12) THEOREM. (Yoshino-Kawamoto [68]) *Let R be an analytic normal local domain of dimension two with residue field k of characteristic 0. Then the following two conditions are equivalent:*

(11.12.1) *The fundamental module E of R is decomposable.*

(11.12.2) *R is an invariant subring of a regular local ring S by a cyclic group G (that is, R is a cyclic quotient singularity).*

PROOF: It suffices to prove (11.12.1) $\Rightarrow$ (11.12.2). So let $E \simeq \mathfrak{a} \oplus \mathfrak{b}$ be a decomposition of the fundamental module of R and let Γ° be the connected component of the AR quiver Γ that contains the class of R. First of all we prove the following claim:

(11.12.3) Under the assumption in (11.12.1) any class of module in Γ° has rank 1.

For this, let M be in Γ°. By definition there is a chain of edges in Γ : $R = M_0 - M_1 - M_2 - \ldots - M_n = M$, where each edge "$-$" indicates either "$\to$" or "$\leftarrow$" in Γ. We prove by induction on the length n of this chain that M has rank 1. If $n = 0$, then there is nothing to prove, for $M \simeq R$. Assume $n > 0$. There might be two possibilities (1) $M_{n-1} \leftarrow M_n$ and (2) $M_{n-1} \to M_n$ in Γ. In the first case, M_n is a direct summand of $\text{Hom}_R(M_{n-1}^*, E)$, hence is isomorphic to either $\text{Hom}_R(M_{n-1}^*, \mathfrak{a})$ or $\text{Hom}_R(M_{n-1}^*, \mathfrak{b})$, both of which are of rank 1 by the induction hypothesis. In the second case, M_{n-1} is isomorphic to one of $\text{Hom}_R(M_n^*, \mathfrak{a})$ and $\text{Hom}_R(M_n^*, \mathfrak{b})$. Since $\text{rank}(M_{n-1}) = 1$ by the induction hypothesis, M_n has also rank 1, so proving (11.12.3).

By virtue of (11.12.3) and Theorem (6.2) one sees that $\mathfrak{C}(R)$ is of finite representation type. Then it is concluded from (11.4) that R is a quotient singularity, that is, there are $S = k\{x, y\}$ and a finite subgroup G of $\text{GL}(2, k)$ such that $R = S^G$. Here, by usual argument using (10.7.1), we may assume that G contains no pseudo-reflections. It remains to prove that G is a cyclic group. Letting F be the $S * G$-module $Sx \oplus Sy$, we showed in (11.8) that the fundamental module E of R is given by F^G. Note from

(11.2) that there is a natural isomorphism of algebras $\text{End}_{S*G}(F) \simeq \text{End}_R(E)$. Since E is decomposable, there is a nontrivial idempotent e in $\text{End}_R(E)$. Considering e as an idempotent in $\text{End}_{S*G}(F)$ by the above isomorphism, F is decomposed into the nontrivial sum $e(F) \oplus (1-e)(F)$ as an $S*G$-module, both summands of which must be free of rank one over S. Thus there is $A = \begin{pmatrix} \alpha & \beta \\ \gamma & \delta \end{pmatrix}$ in $\text{GL}(2,k)$ such that $e(F) = S(\alpha x + \beta y)$ and $(1-e)(F) = S(\gamma x + \delta y)$. After changing variables by the linear transformation A, we may assume that $e(F) = Sx$ and $(1-e)(F) = Sy$. Therefore the action of G is diagonalized. Let χ be the character of G given by $\chi(g) = a$ for $g = \begin{pmatrix} a & 0 \\ 0 & b \end{pmatrix} \in G$. Since G contains no pseudo-reflections, this χ gives an injective homomorphism from G into k^*. Thus G is a subgroup of k^*, hence it is cyclic and the proof of the theorem is completed. ∎

(11.13) REMARK. The above proof shows how to describe the AR quiver for a cyclic quotient singularity. Actually it follows from (11.12.3) and (6.2) that any indecomposable CM module has rank one, so there is a one-to-one correspondence between the elements in the class group $\text{Cl}(R)$ of R and the vertices in the AR quiver Γ for R. Suppose the fundamental module E is decomposed as the sum of ideals $\mathfrak{a}$ and $\mathfrak{b}$, and denote by a, b the class of $\mathfrak{a}$, $\mathfrak{b}$ in $\text{Cl}(R)$. Then the AR quiver Γ is figured as follows:

(11.13.1)

$$\begin{array}{ccccccccccc}
3a+b & & 2a & & a-b & & -2b & & \\
& 2a+b & & a & & -b & & a-2b & \\
2a+2b & & a+b & & 0 & & -a-b & & \\
& a+2b & & b & & -a & & -2a-b & \\
a+3b & & 2b & & -a+b & & -2a & &
\end{array}$$

where $na + mb$ denotes the ideal class of $((\mathfrak{a}^n\mathfrak{b}^m)^{-1})^{-1}$ in $\text{Cl}(R)$. Since this diagram must be finite by (11.12), some vertices in (11.13.1) should be identified. For example, if a has order n in $\text{Cl}(R)$, then $0 = na = 2na = \ldots$ so on. The AR quiver for R is thus obtained from the graph (11.13.1) by dividing it by the relations in the divisor class group.

Chapter 12. Knörrer's periodicity

In Chapter 8 we showed that a local ring of hypersurface is a simple singularity when it is of finite representation type. Conversely simple singularities of dimension 1 and 2 are of finite representation type as we have shown in Chapter 9 and 10 respectively. One of our aims here is to show that the latter is true as well in arbitrary dimension, so that, for hypersurfaces, the notion of simple singularity is equivalent to that of finite representation type.

Throughout this chapter k is an *algebraically closed field of characteristic* 0 and S is the power series ring over k in $d+1$ variables:

$$S = k\{x, y, z_2, z_3, \ldots, z_d\}.$$

Set $R = S/(f)$ where f is one of the following polynomials:

(A_n) $x^2 + y^{n+1} + z_2^2 + z_3^2 + \ldots + z_d^2$ $(n \geq 1)$,

(D_n) $x^2 y + y^{n-1} + z_2^2 + z_3^2 + \ldots + z_d^2$ $(n \geq 4)$,

(E_6) $x^3 + y^4 + z_2^2 + z_3^2 + \ldots + z_d^2$,

(E_7) $x^3 + xy^3 + z_2^2 + z_3^2 + \ldots + z_d^2$,

(E_8) $x^3 + y^5 + z_2^2 + z_3^2 + \ldots + z_d^2$.

Recall that these are equations of simple singularities; (8.8). We want to show below that R is of finite representation type. Since we have already shown this for the cases when $d = 1$ and 2, the main idea here is to reduce the problem to one in lower dimension. For this purpose, consider new rings as follows:

(12.1.1) $S^\sharp = S\{u\} = k\{x, y, z_2, z_3, \ldots, z_d, u\}$,
                $R^\sharp = S^\sharp/(f + u^2)$,

with u a new variable. Note that, if R has one of the types (A), (D) and (E), then $R^\sharp$ has the same type as R. Also note that the dimension $R^\sharp$ is $(d+1)$ while R has dimension d. Clearly we have $R = R^\sharp/(u)$, so that any R-module can be regarded as an $R^\sharp$-module. As before we shall denote by $\mathfrak{C}(R)$ and $\mathfrak{C}(R^\sharp)$ the categories of CM modules over R and $R^\sharp$ respectively.

(12.1) DEFINITION. Let N be in $\mathfrak{C}(R^\sharp)$. We define an R-module $\Xi(N)$ by putting $\Xi(N) = N/uN$. Since $\text{depth}(N/uN) = \text{depth}(N) - 1 = \dim(R)$, we have $\Xi(N) \in \mathfrak{C}(R)$. Similarly for a morphism φ in $\mathfrak{C}(R^\sharp)$ we put $\Xi(\varphi) = \varphi \otimes R^\sharp/(u)$. Thus we have defined the functor Ξ from $\mathfrak{C}(R^\sharp)$ to $\mathfrak{C}(R)$.

Note that, by (12.1.1), S is a subring of $R^\sharp$ on which $R^\sharp$ is generated by 1 and u, i.e. $S \subset R^\sharp$ is a Noetherian normalization. Hence an $R^\sharp$-module N is CM if and only if it is free as an S-module, (1.9). In other words, CM $R^\sharp$-modules are the free S-modules with $R^\sharp$-module structure, or equivalently, the free S-modules on which u acts. Notice that the action of u must satisfy $u^2 = -f$. Therefore there is a one-to-one correspondence between the set of CM modules over $R^\sharp$ and the set of square matrices φ with entries in S and with the property $\varphi^2 = -f \cdot I$. Clearly isomorphic modules correspond to equivalent matrices, where we say that matrices φ and ψ are equivalent if $\alpha \varphi \alpha^{-1} = \psi$ for some invertible matrix α.

First of all we remark:

(12.2) LEMMA. *Let N be a CM module over $R^\sharp$ and let φ be an $(n \times n)$-matrix over S given by the action of u on N. Then $(uI_n - \varphi, uI_n + \varphi)$ lies in $\text{MF}_{S^\sharp}(f + u^2)$ with $\text{Coker}(uI_n - \varphi, uI_n + \varphi) = N$. Similarly $(\varphi, -\varphi)$ is in $\text{MF}_S(f)$ such that $\text{Coker}(\varphi, -\varphi) = \Xi(N)$.*

PROOF: It is easy to check that $\varphi \cdot (-\varphi) = (-\varphi) \cdot \varphi = f \cdot I_n$ and $(uI_n - \varphi) \cdot (uI_n + \varphi) = (uI_n + \varphi) \cdot (uI_n - \varphi) = (f + u^2) \cdot I_n$. Hence they give matrix factorizations of f and $f + u^2$ respectively. To prove the lemma we have to show that $N \simeq \text{Coker}(R^{\sharp(n)} \overset{uI_n - \varphi}{\to} R^{\sharp(n)})$ and $\Xi(N) \simeq \text{Coker}(R^{(n)} \overset{\varphi}{\to} R^{(n)})$. For the first equality, let N' be the module of the right hand side and let $\{e_1, e_2, \ldots, e_n\}$ be a free base of $R^{\sharp(n)}$. Then note that, as an S-module, $R^{\sharp(n)}$ has a free base $\{e_1, e_2, \ldots, e_n, ue_1, ue_2, \ldots, ue_n\}$. Since $(uI - \varphi)(e_i) = -\varphi(e_i) + ue_i$ and $(uI - \varphi)(ue_i) = fe_i - u\varphi(e_i)$, we have

$$N' \simeq \text{Coker}(\begin{pmatrix} -\varphi & fI \\ I & \varphi \end{pmatrix} : S^{(2n)} \longrightarrow S^{(2n)})$$

as an S-module, so that N' is an S-free module having the image of $\{e_1, e_2, \ldots, e_n\}$ as a base, on which u acts as the matrix φ. Hence by the choice of φ, N' is isomorphic to N

as an $R^\sharp$-module. This proves the first equality. The second equality is immediate from the first. In fact,

$$\Xi(N) = N \otimes_{R^\sharp} R \simeq \operatorname{Coker}(R^{\sharp(n)} \xrightarrow{\varphi - uI} R^{\sharp(n)}) \otimes_{R^\sharp} R \simeq \operatorname{Coker}(R^{(n)} \xrightarrow{\varphi} R^{(n)}). \qquad \blacksquare$$

Recall that, for a module M in $\mathfrak{C}(R)$, $\operatorname{syz}^1_{R^\sharp}(M)$ is the reduced first syzygy of M as an $R^\sharp$-module, i.e. there is an exact sequence

$$0 \longrightarrow \operatorname{syz}^1_{R^\sharp}(M) \oplus F \longrightarrow G \longrightarrow M \longrightarrow 0,$$

where F and G are free over $R^\sharp$, cf. (1.15). From this we have

$$\operatorname{depth}(\operatorname{syz}^1_{R^\sharp}(M)) = \operatorname{depth}(M) + 1 = \dim(R) + 1 = \dim(R^\sharp),$$

in particular, $\operatorname{syz}^1_{R^\sharp}(M)$ is a CM module over $R^\sharp$. We can prove the following:

(12.3) LEMMA. *Let M be in $\mathfrak{C}(R)$ and let (φ, ψ) be the matrix factorization of f with $M = \operatorname{Coker}(\varphi, \psi)$. Suppose that M has no free summands and that (φ, ψ) is chosen to be reduced. Then $(\begin{pmatrix} \psi & -uI \\ uI & \varphi \end{pmatrix}, \begin{pmatrix} \varphi & uI \\ -uI & \psi \end{pmatrix})$ is an element in $\operatorname{MF}_{S^\sharp}(f + u^2)$ such that*

$$\operatorname{Coker}(\begin{pmatrix} \psi & -uI \\ uI & \varphi \end{pmatrix}, \begin{pmatrix} \varphi & uI \\ -uI & \psi \end{pmatrix}) = \operatorname{syz}^1_{R^\sharp}(M).$$

PROOF: Since $R = R^\sharp/(u)$, we have a commutative diagram:

(12.3.1)
$$\begin{array}{ccccccccc}
& & 0 & & 0 & & 0 & & \\
& & \uparrow & & \uparrow & & \uparrow & & \\
& & R^{(n)} & \xrightarrow{\psi} & R^{(n)} & \xrightarrow{\varphi} & R^{(n)} & \longrightarrow & M & \longrightarrow 0 \\
& & \uparrow & & \uparrow & & \uparrow & & \\
& & R^{\sharp(n)} & \xrightarrow{\psi} & R^{\sharp(n)} & \xrightarrow{\varphi} & R^{\sharp(n)} & & \\
& & u\uparrow & & u\uparrow & & u\uparrow & & \\
& & R^{\sharp(n)} & \xrightarrow{\psi} & R^{\sharp(n)} & \xrightarrow{\varphi} & R^{\sharp(n)}, & &
\end{array}$$

where the columns and the top row are exact, while the second and the third rows are not even complexes. However, by chasing the diagram, we get a free presentation of M

as an $R^\sharp$-module:

(12.3.2)
$$R^{\sharp(2n)} \xrightarrow{\begin{pmatrix} \psi & -uI \\ uI & \varphi \end{pmatrix}} R^{\sharp(2n)} \xrightarrow{(\varphi, uI)} R^{\sharp(n)} \longrightarrow M \longrightarrow 0.$$

In fact, it is easy from (12.3.1) to see that there is an exact sequence $R^{\sharp(2n)} \xrightarrow{(\varphi,uI)} R^{\sharp(n)} \to M \to 0$. Assume that $\begin{pmatrix} \alpha \\ \beta \end{pmatrix}$ is in $(R^{\sharp(n)})^{(2)}$ with $\varphi(\alpha) + u\beta = 0$. Then $\varphi(\alpha \mod (u)) = \varphi(\alpha) \mod (u) = 0$, hence by (12.3.1) there are γ and δ in $R^{\sharp(n)}$ such that $\alpha = \psi(\gamma) - u\delta$. From this we have $u\beta = -\varphi\psi(\gamma) + \varphi(u\delta) = u^2\gamma + u\varphi(\delta)$, since $\varphi\psi = fI = -u^2I$. Thus $\beta = u\gamma + \varphi(\delta)$, so that $\begin{pmatrix} \alpha \\ \beta \end{pmatrix} = \begin{pmatrix} \psi & -uI \\ uI & \varphi \end{pmatrix} \begin{pmatrix} \gamma \\ \delta \end{pmatrix}$. This proves the exactness of (12.3.2). We therefore have $\mathrm{syz}^1_{R^\sharp}(M) \simeq \mathrm{Coker}\begin{pmatrix} \psi & -uI \\ uI & \varphi \end{pmatrix}$ up to free summands. On the other hand, as the pair of matrices $(\begin{pmatrix} \psi & -uI \\ uI & \varphi \end{pmatrix}, \begin{pmatrix} \varphi & uI \\ -uI & \psi \end{pmatrix})$ is a reduced matrix factorization of $f + u^2$, this matrix factorization must correspond to $\mathrm{syz}^1_{R^\sharp}(M)$ by (7.7). ∎

The above lemmas amount to showing the following:

(12.4) PROPOSITION. *Let $M \in \mathfrak{C}(R)$ and let $N \in \mathfrak{C}(R^\sharp)$. Suppose M has no free summands. Then*

(12.4.1) $\qquad\qquad \Xi(\mathrm{syz}^1_{R^\sharp}(M)) \simeq M \oplus \mathrm{syz}^1_R(M),$

(12.4.2) $\qquad\qquad \mathrm{syz}^1_{R^\sharp}(\Xi(N)) \simeq N \oplus \mathrm{syz}^1_{R^\sharp}(N).$

PROOF: (12.4.1): Let (φ, ψ) be a reduced matrix factorization in $\mathrm{MF}_S(f)$ with $\mathrm{Coker}(\varphi, \psi) = M$. Then we know from the previous lemma that $\mathrm{syz}^1_{R^\sharp}(M) = \mathrm{Coker}(\begin{pmatrix} \psi & -uI \\ uI & \varphi \end{pmatrix}, \begin{pmatrix} \varphi & uI \\ -uI & \psi \end{pmatrix})$. Hence we have isomorphisms

$$\Xi(\mathrm{syz}^1_{R^\sharp}(M)) \simeq \mathrm{Coker}(\begin{pmatrix} \psi & -uI \\ uI & \varphi \end{pmatrix} \otimes R^\sharp/(u))$$
$$\simeq \mathrm{Coker}(R^{(n)} \xrightarrow{\varphi} R^{(n)}) \oplus \mathrm{Coker}(R^{(n)} \xrightarrow{\psi} R^{(n)})$$
$$\simeq M \oplus \mathrm{syz}^1_R(M).$$

(12.4.2): Let φ be the matrix on S defined as in (12.2). Then $(\varphi, -\varphi)$ is in $\mathrm{MF}_S(f)$ with $\mathrm{Coker}(\varphi, -\varphi) = \Xi(N)$ by (12.2). Hence it follows from (12.3) that $\mathrm{syz}^1_{R^\sharp}(\Xi(M)) =$

$\operatorname{Coker}(\begin{pmatrix} -\varphi & -uI \\ uI & \varphi \end{pmatrix}, \begin{pmatrix} \varphi & uI \\ -uI & -\varphi \end{pmatrix})$. Here note that

$$\frac{1}{2}\begin{pmatrix} 1 & 1 \\ 1 & -1 \end{pmatrix}\begin{pmatrix} -\varphi & -uI \\ uI & \varphi \end{pmatrix}\begin{pmatrix} 1 & 1 \\ 1 & -1 \end{pmatrix} = \begin{pmatrix} 0 & uI - \varphi \\ -uI - \varphi & 0 \end{pmatrix}.$$

Hence the above becomes

$$\operatorname{syz}^1_{R^\sharp}(\Xi(N)) \simeq \operatorname{Coker}(uI - \varphi, uI + \varphi) \oplus \operatorname{Coker}(uI + \varphi, uI - \varphi) \simeq N \oplus \operatorname{syz}^1_{R^\sharp}(N)$$

by (12.2). ∎

Now we are ready to prove the main result of this chapter.

(12.5) THEOREM. (Knörrer [44]) $R^\sharp$ *is of finite representation type if and only if R is.*

Taking into account (9.3) and (10.15) together with this theorem, we have a striking result due to Knörrer.

(12.6) COROLLARY. *Simple singularities of any dimension are of finite representation type.*

PROOF OF THE THEOREM: Let R be of finite representation type, and suppose that $\{M_1, M_2, \ldots, M_m\}$ is the complete set of nonisomorphic indecomposable CM modules over R. If N is an object in $\mathfrak{C}(R^\sharp)$, then $\Xi(N)$ lies in $\mathfrak{C}(R)$ by (12.1), hence we may write $\Xi(N) = \sum_i M_i^{(n_i)}$. Then by (12.4.2), $\sum_i \operatorname{syz}^1_{R^\sharp}(M_i)^{(n_i)} \simeq \operatorname{syz}^1_{R^\sharp}(\Xi(N)) \simeq N \oplus \operatorname{syz}^1_{R^\sharp}(N)$ as an $R^\sharp$-module. If N is indecomposable, then this implies that N is isomorphic to a direct summand of some $\operatorname{syz}^1_{R^\sharp}(M_i)$, hence such classes of N are finite.

Conversely suppose that $R^\sharp$ is of finite representation type and that $\{N_1, N_2, \ldots, N_n\}$ is the complete set of nonisomorphic indecomposable CM modules over $R^\sharp$. Given an indecomposable CM module M over R, we know that $\operatorname{syz}^1_{R^\sharp}(M)$ is in $\mathfrak{C}(R^\sharp)$, hence we may write $\operatorname{syz}^1_{R^\sharp}(M) \simeq \sum_i N_i^{(m_i)}$. If M is not free, then by (12.4.1) we see that $\sum_i \Xi(N_i)^{(m_i)} \simeq \Xi(\operatorname{syz}^1_{R^\sharp}(M)) \simeq M \oplus \operatorname{syz}^1_R(M)$. Therefore M is a summand of some $\Xi(N_i)$. Hence the classes of nonfree indecomposable CM modules over R are finite. ∎

(12.7) REMARK. By this proof we get a simple way to construct CM modules over $R^\sharp$ from those on R. Let the ring R and all indecomposable CM modules $\{M_i\}$ over R be given. Then one obtains all indecomposable, non-free modules over $R^\sharp$ as summands of $\operatorname{syz}^1_{R^\sharp}(M_i)$. Here note that each $\operatorname{syz}^1_{R^\sharp}(M_i)$ is decomposed into at most two modules, since $\Xi(\operatorname{syz}^1_{R^\sharp}(M_i))$ has only two direct summands by (12.4).

However it is often difficult to perform the actual computation along this line. The situation will be better if we take the $\sharp$-operation twice.

(12.8) DEFINITION. According to (12.1.1) we define:

$$R^{\sharp\sharp} = S\{u,v\}/(f + u^2 + v^2),$$

where u and v are variables on S. Let (φ, ψ) be in $\mathrm{MF}_S(f)$. Then define a pair of matrices over $S\{u, v\}$ by

(12.8.1) $$\Omega(\varphi, \psi) = (\begin{pmatrix} \varphi & \xi \\ \eta & -\psi \end{pmatrix}, \begin{pmatrix} \psi & \xi \\ \eta & -\varphi \end{pmatrix}),$$

where $\xi = u + \sqrt{-1}v$ and $\eta = u - \sqrt{-1}v$. It is immediate that $\Omega(\varphi, \psi)$ is a matrix factorization of $f + u^2 + v^2$, and thus

$$\Omega(\varphi, \psi) \in \mathrm{MF}_{S\{u,v\}}(f + u^2 + v^2).$$

For a morphism $(\alpha, \beta) : (\varphi, \psi) \to (\varphi', \psi')$ in $\mathrm{MF}_S(f)$, we define $\Omega(\alpha, \beta)$ to be a morphism

$$(\begin{pmatrix} \alpha & 0 \\ 0 & \beta \end{pmatrix}, \begin{pmatrix} \beta & 0 \\ 0 & \alpha \end{pmatrix}) : \Omega(\varphi, \psi) \longrightarrow \Omega(\varphi', \psi')$$

in $\mathrm{MF}_{S\{u,v\}}(f + u^2 + v^2)$. It is quite elementary to show that $\Omega(\alpha, \beta)$ is well-defined and that we have a functor $\Omega : \mathrm{MF}_S(f) \to \mathrm{MF}_{S\{u,v\}}(f + u^2 + v^2)$.

If (φ, ψ) is a reduced matrix factorization, then $\Omega(\varphi, \psi)$ is also reduced. Similarly, $\Omega(f, 1)$ and $\Omega(1, f)$ are direct sums of $(f + u^2 + v^2, 1)$ and $(1, f + u^2 + v^2)$. Hence we can define a functor:

$$\underline{\mathrm{RMF}}_S(f) \longrightarrow \underline{\mathrm{RMF}}_{S\{u,v\}}(f + u^2 + v^2),$$

which we also denote by Ω, cf. (7.3). Thus, by virtue of (7.4), we have a functor

$$\underline{\mathfrak{C}}(R) \longrightarrow \underline{\mathfrak{C}}(R^{\sharp\sharp}),$$

which is also denoted by Ω.

Note that, for a reduced matrix factorization (φ, ψ) in $\underline{\mathrm{RMF}}_S(f)$,

$$\Omega(\mathrm{Coker}(\varphi, \psi)) = \mathrm{Coker}(\Omega(\varphi, \psi)),$$

in $\underline{\mathfrak{C}}(R^{\sharp\sharp})$. Moreover, by (7.7) and (12.8.1), we have

$$\mathrm{Coker}(\Omega(\psi, \varphi)) = \mathrm{syz}^1_{R^{\sharp\sharp}}(\mathrm{Coker}(\Omega(\varphi, \psi))).$$

We note the following:

(12.9)

LEMMA. Let $(\begin{pmatrix} A & B \\ C & D \end{pmatrix}, \begin{pmatrix} A' & B' \\ C' & D' \end{pmatrix})$ be a morphism from $(\begin{pmatrix} \varphi & \xi \\ \eta & -\psi \end{pmatrix}, \begin{pmatrix} \psi & \xi \\ \eta & -\varphi \end{pmatrix})$ to $(\begin{pmatrix} \varphi' & \xi \\ \eta & -\psi' \end{pmatrix}, \begin{pmatrix} \psi' & \xi \\ \eta & -\varphi' \end{pmatrix})$ in $\mathrm{MF}_{S\{u,v\}}(f + u^2 + v^2)$. Assume that all the entries in A' are in the ideal $(\xi, \eta) S\{u, v\}$, then the following extension splits:

$$\begin{pmatrix} \psi' & \xi & A' & B' \\ \eta & -\varphi' & C' & D' \\ & & \varphi & \xi \\ & & \eta & -\psi \end{pmatrix},$$

cf. (7.8).

PROOF: Since we can write $A' = A'_1 \xi + A'_2 \eta$, applying an elementary transformation of matrices, we may assume that $A' = 0$. By the definition of morphisms of matrix factorizations, we have:

$$\begin{pmatrix} \varphi' & \xi \\ \eta & -\psi' \end{pmatrix} \begin{pmatrix} 0 & B' \\ C' & D' \end{pmatrix} = \begin{pmatrix} A & B \\ C & D \end{pmatrix} \begin{pmatrix} \varphi & \xi \\ \eta & -\psi \end{pmatrix}$$

$$\begin{pmatrix} 0 & B' \\ C' & D' \end{pmatrix} \begin{pmatrix} \psi & \xi \\ \eta & -\varphi \end{pmatrix} = \begin{pmatrix} \psi' & \xi \\ \eta & -\varphi' \end{pmatrix} \begin{pmatrix} A & B \\ C & D \end{pmatrix}$$

In particular,

(i) $\eta B' = \psi' A + \xi C$
(ii) $\xi C' = A\varphi + \eta B$
(iii) $\varphi' B' + \xi D' = \xi A - B\psi$

Multiplying (i) by φ' from the left and using $\varphi'\psi' = fI$, we get $\eta \varphi' B' = fA + \xi \varphi' C$. Since $\{f, \xi, \eta\}$ is a regular sequence on $S\{u, v\}$, we may write

(iv) $\qquad A = \xi P + \eta Q,$

for some matrices P, Q on $S\{u, v\}$. Substitute (iv) into (i) to get

$$\xi(\psi' P + C) = \eta(B' - \psi' Q),$$

and thus there is a matrix P' such that

$$B' - \psi' Q = \xi P', \quad \psi' P + C = \eta P'.$$

Similarly we subsitute (iv) into (ii) and deduce the existence of a matrix Q' on $S\{u,v\}$ with
$$C' - P\varphi = \eta Q', \quad Q\varphi + B = \xi Q'.$$
Then we have the equality of matrices:

$$\begin{pmatrix} 1 & & & \\ 1 & -P & -Q' & \\ & & 1 & \\ & & & 1 \end{pmatrix} \begin{pmatrix} \psi' & \xi & 0 & B' \\ \eta & -\varphi' & C' & D' \\ & & \varphi & \xi \\ & & \eta & -\psi \end{pmatrix} \begin{pmatrix} 1 & & -Q & \\ & 1 & & -P' \\ & & 1 & \\ & & & 1 \end{pmatrix} = \begin{pmatrix} \psi' & \xi & 0 & E \\ \eta & -\varphi' & F & G \\ & & \varphi & \xi \\ & & \eta & -\psi \end{pmatrix},$$

where
$$E = B' - \psi'Q - \xi P' = 0, \qquad F = C' - P\varphi - Q'\eta = 0,$$
and
$$G = D' - P\xi + Q'\psi - \eta Q + \varphi'P' = D' - A + Q'\psi + \varphi'P'.$$

Since
$$\xi G = \xi D' - \xi A + B\psi + \varphi'B' = 0 \quad \text{by } (iii)$$
and since ξ is a nonzero divisor on $S\{u,v\}$, we see that $G = 0$ and thus the extension is equivalent to the split one. ∎

(12.10) THEOREM. (Knörrer [44]) *The functor Ω defined in (12.8) gives the equivalence of categories:*
$$\underline{\mathfrak{C}}(R) \simeq \underline{\mathfrak{C}}(R^{\sharp\sharp}).$$

PROOF: First of all we show that Ω is fully faithful. To do this, take any objects (φ, ψ), (φ', ψ') in $\underline{\text{RMF}}_S(f)$. Putting $M = \text{Coker}(\varphi, \psi)$ and $M' = \text{Coker}(\varphi', \psi')$, we consider a homomorphism of Abelian groups induced by Ω:

$$\rho : \underline{\text{Hom}}_R(M, M') \to \underline{\text{Hom}}_{R^{\sharp\sharp}}(\Omega(M), \Omega(M')).$$

We have to show that ρ is bijective.

Considering the exact sequence $0 \to \text{syz}_R^1(M') \to F \to M' \to 0$ with F a free R-module, we see that the sequence

$$\text{Hom}_R(M, F) \longrightarrow \text{Hom}_R(M, M') \longrightarrow \text{Ext}_R^1(M, \text{syz}_R^1 M') \longrightarrow 0$$

is exact because $\mathrm{Ext}^1_R(M, R) = 0$. Then it is clear from the definition of $\underline{\mathrm{Hom}}$ in (3.7) that there is an R-isomorphism

$$\mu : \underline{\mathrm{Hom}}_R(M, M') \to \mathrm{Ext}^1_R(M, \mathrm{syz}^1_R M').$$

The mapping μ is realized as follows: For any $h \in \underline{\mathrm{Hom}}_R(M, M')$ it can be written as $\mathrm{Coker}(\alpha, \beta)$ for some morphism $(\alpha, \beta) : (\varphi, \psi) \to (\varphi', \psi')$. Then $\mu(h)$ corresponds to the extension of M by $\mathrm{syz}^1_R M'$ which is given by the matrix $\begin{pmatrix} \psi' & \beta \\ 0 & \varphi \end{pmatrix}$; see (7.8). Likewise, we have an $R^{\sharp\sharp}$-isomorphism

$$\nu : \underline{\mathrm{Hom}}_{R^{\sharp\sharp}}(\Omega(M), \Omega(M')) \to \mathrm{Ext}^1_{R^{\sharp\sharp}}(\Omega(M), \Omega(\mathrm{syz}^1_R M'))$$

and we see that, for any $h = \mathrm{Coker}(\alpha, \beta)$ in $\underline{\mathrm{Hom}}_R(M, M')$, $\nu\rho(h)$ corresponds to the extension

(12.10.1)
$$E = \begin{pmatrix} \psi' & \xi & \beta & 0 \\ \eta & -\varphi' & 0 & \alpha \\ & & \varphi & \xi \\ & & \eta & -\psi \end{pmatrix}.$$

Now we prove that ρ is bijective. Let $h = \mathrm{Coker}(\alpha, \beta)$ be in $\underline{\mathrm{Hom}}_R(M, M')$ and suppose that $\rho(h) = 0$. Then the extension E in (12.10.1) is split, or equivalently, there are matrices A, B on $S\{u, v\}$ with

$$\begin{pmatrix} 1 & A \\ 0 & 1 \end{pmatrix} E \begin{pmatrix} 1 & B \\ 0 & 1 \end{pmatrix} = \begin{pmatrix} \psi' & \xi & 0 & 0 \\ \eta & -\varphi' & 0 & 0 \\ 0 & 0 & \varphi & \xi \\ 0 & 0 & \eta & -\psi \end{pmatrix}.$$

Substituting $u = v = 0$, we see that this equation becomes $\begin{pmatrix} \psi' & \beta \\ 0 & \varphi \end{pmatrix} \oplus \begin{pmatrix} -\varphi' & \alpha \\ 0 & -\psi \end{pmatrix} \simeq (\psi') \oplus (\varphi') \oplus (\varphi) \oplus (\psi)$, hence that the extension $\begin{pmatrix} \psi' & \beta \\ 0 & \varphi \end{pmatrix}$ is also split. Thus $\mu(h) = 0$ and hence $h = 0$. This shows the injectivity of ρ.

To prove the surjectivity, take any element $h = \mathrm{Coker}(\begin{pmatrix} A & B \\ C & D \end{pmatrix}, \begin{pmatrix} A' & B' \\ C' & D' \end{pmatrix})$ in $\underline{\mathrm{Hom}}_{R^{\sharp\sharp}}(\Omega(M), \Omega(M'))$. Letting $A_0 = A|_{u=v=0}$ and $A'_0 = A'|_{u=v=0}$, we see from the equation $\begin{pmatrix} A & B \\ C & D \end{pmatrix} \begin{pmatrix} \varphi & \xi \\ \eta & -\psi \end{pmatrix} = \begin{pmatrix} \varphi' & \xi \\ \eta & -\psi' \end{pmatrix} \begin{pmatrix} A' & B' \\ C' & D' \end{pmatrix}$ that $A_0 \varphi = \varphi' A'_0$ and so that (A_0, A_0') is a morphism of matrix factorizations $(\varphi, \psi) \to (\varphi', \psi')$. Let $a = \mathrm{Coker}(A_0, A_0') \in \mathrm{Hom}_R(M, M')$. It is then immediately clear that

$$h - \rho(a) = \mathrm{Coker}\left(\begin{pmatrix} A - A_0 & B \\ C & D - A_0' \end{pmatrix}, \begin{pmatrix} A' - A_0' & B' \\ C' & D' - A_0 \end{pmatrix}\right)$$

and that $\nu(h - \rho(a))$ is the extension

$$\begin{pmatrix} \psi' & \xi & A' - A_0' & B' \\ \eta & -\varphi' & C' & D' - A_0 \\ & & \varphi & \xi \\ & & \eta & -\psi \end{pmatrix}.$$

Since any entries in $A' - A_0'$ are in $(\xi, \eta)S\{u, v\}$, we conclude from (12.9) that $\nu(h - \rho(a)) = 0$. Therefore $h = \rho(a)$, proving that ρ is surjective. Thus we have shown that the functor Ω is fully faithful.

It remains to be shown that Ω gives a surjective mapping onto the set of objects of $\underline{\mathfrak{C}}(R^{\#\#})$. To do this, it is sufficient to prove that any indecomposable object N in $\underline{\mathfrak{C}}(R^{\#\#})$ has the form $\Omega(M)$ for some $M \in \underline{\mathfrak{C}}(R)$. Note that N is a direct summand of $\mathrm{syz}^1_{R^{\#\#}}(\mathrm{syz}^1_{R^\#}(M))$ for some indecomposable nonfree CM module M over R. (Use Remark (12.7) twice.) If (φ, ψ) is the matrix factorization of f corresponding to M, then we see by using (12.3) twice that the matrix factorization of $f + u^2 + v^2$ corresponding to $\mathrm{syz}^1_{R^{\#\#}}(\mathrm{syz}^1_{R^\#}(M))$ is given by

$$\left(\begin{pmatrix} \varphi & uI & -vI & 0 \\ -uI & \psi & 0 & -vI \\ vI & 0 & \psi & -uI \\ 0 & vI & uI & \varphi \end{pmatrix}, \begin{pmatrix} \psi & -uI & vI & 0 \\ uI & \varphi & 0 & vI \\ -vI & 0 & \varphi & uI \\ 0 & -vI & -uI & \psi \end{pmatrix}\right).$$

An easy computation shows that this is equvalent to the following matrix factorization:

$$\left(\begin{pmatrix} \psi & \xi & 0 & 0 \\ \eta & -\varphi & 0 & 0 \\ 0 & 0 & \varphi & \xi \\ 0 & 0 & \eta & -\psi \end{pmatrix}, \begin{pmatrix} \varphi & \xi & 0 & 0 \\ \eta & -\psi & 0 & 0 \\ 0 & 0 & \psi & \xi \\ 0 & 0 & \eta & -\varphi \end{pmatrix}\right).$$

Hence we have an isomorphism of $R^{\#\#}$-modules:

$$\mathrm{syz}^1_{R^{\#\#}}(\mathrm{syz}^1_{R^\#}(M)) \simeq \Omega(M) \oplus \Omega(\mathrm{syz}^1_R M).$$

Therefore N is a direct summand of this module, hence of $\Omega(M)$ or $\Omega(\operatorname{syz}_R^1 M)$. In either case N is in the image of Ω, and the proof is finished. ∎

By the equivalence in Theorem (12.10) we can deduce various properties of CM modules on $R^\sharp$ from those on R. For example we have:

(12.11) COROLLARY. *Let M and N be non-free indecomposable CM modules over R and let g be an R-homomorphism from M to N. Then;*
(12.11.1) *$\Omega(M)$ is an indecomposable $R^\sharp$-module.*
(12.11.2) *g is a split monomorphism (resp. a split epimorphism) if and only if $\Omega(g)$ is a split monomorphism (resp. a split epimorphism).*
(12.11.3) *g is an irreducible morphism from M to N if and only if $\Omega(g)$ is an irreducible morphism from $\Omega(M)$ to $\Omega(N)$.*

PROOF: (12.11.1) : By the equivalence of categories in (12.10), $\Omega(M)$ is indecomposable as an object in $\underline{\mathfrak{C}}(R^\sharp)$. Therefore if $\Omega(M)$ is decomposable as an $R^\sharp$-module, the only possibility is that $\Omega(M)$ has a free summand. However this is not the case, since the matrix factorization $\Omega(\varphi, \psi)$ is reduced if (φ, ψ) is the reduced matrix factorization corresponding to M; cf. (7.5.1).

(12.11.2): We prove this only for the case of split monomorphisms.
If g is a split monomorphism, then it is obvious that $\Omega(g)$ is also a split monomorphism. Suppose $\Omega(g)$ is a split monomorphism. Then there is a homomorphism $\pi \in \operatorname{Hom}_{R^\sharp}(\Omega(N), \Omega(M))$ with $\pi \cdot \Omega(g) = 1_{\Omega(M)}$. It follows from (12.10) that there is an $h \in \operatorname{Hom}_R(N, M)$ such that the image of $\Omega(h) - \pi$ in $\underline{\operatorname{Hom}}_{R^\sharp}(\Omega(N), \Omega(M))$ is trivial. It then turns out that $\Omega(h \cdot g - 1_M)$ has a trivial image in $\underline{\operatorname{End}}_{R^\sharp}(\Omega(M))$, and hence $h \cdot g = 1_M$ as an element of $\underline{\operatorname{End}}_R(M)$. Since $\operatorname{End}_R(M)$ is local and since $\underline{\operatorname{End}}_R(M)$ is a homomorphic image of $\operatorname{End}_R(M)$, we see that $h \cdot g$ is an automorphism of M and hence g is a split monomorphism.

(12.11.3): Suppose g is irreducible and that we are given a commutative diagram in $\underline{\mathfrak{C}}(R^\sharp)$:

$$\begin{array}{ccc} \Omega(M) & \xrightarrow{\Omega(g)} & \Omega(N) \\ {}_\mu \searrow & & \nearrow {}_\nu \\ & Q & \end{array}$$

Then by (12.10) we have a CM module L over R and $h \in \operatorname{Hom}_R(M, L)$, $k \in \operatorname{Hom}_R(L, N)$ with $\mu = \Omega(h)$ in $\underline{\operatorname{Hom}}_{R^\sharp}(\Omega(M), Q)$ and $\nu = \Omega(k)$ in $\underline{\operatorname{Hom}}_{R^\sharp}(Q, \Omega(N))$. Since $g - k \cdot h$ is trivial as an element of $\underline{\operatorname{Hom}}_R(M, N)$, there are a free R-module F, $a \in \operatorname{Hom}_R(M, F)$ and $b \in \operatorname{Hom}_R(F, N)$ such that $g = k \cdot h + b \cdot a$ as an element of $\operatorname{Hom}_R(M, N)$. Because g is irreducible, this shows that either $\begin{pmatrix} h \\ a \end{pmatrix}$ is a split monomorphism or (k, b) is a split epimorphism. Note that neither a nor b are split morphisms, since M and N are non-free and indecomposable. Thus by (1.21) and by its dual statement we see that either h is a split monomorphism or k is a split epimorphism. Hence, by (12.11.2), either μ is a split monomorphism or ν is a split epimorphism. This shows that $\Omega(g)$ is irreducible; cf. (2.10).

The converse is proved in a similar way and we leave it to the reader. ∎

Let $\Gamma(R)$ (resp. $\Gamma(R^{\sharp\sharp})$) be the AR quiver of R (resp. $R^{\sharp\sharp}$) and define $\underline{\Gamma}(R)$ (resp. $\underline{\Gamma}(R^{\sharp\sharp})$) to be the graph obtained from $\Gamma(R)$ (resp. $\Gamma(R^{\sharp\sharp})$) by deleting the vertex of indecomposable free module and any arrows connecting with this vertex. $\underline{\Gamma}(R)$ is called the **stable AR quiver** of R. Then (12.11.3) implies the following:

(12.12) COROLLARY.

$$\underline{\Gamma}(R) = \underline{\Gamma}(R^{\sharp\sharp})$$

(12.13) REMARK. It is known that the number of arrows ending in (or starting from) the vertex of free module is doubled when passing from $\Gamma(R)$ to $\Gamma(R^{\sharp\sharp})$. See Knörrer [**44**] or Solberg [**60**] for further discussion.

Chapter 13. Grothendieck groups

We show in this chapter that the Grothendieck group of the category of CM modules can be computed from the AR quiver if it is finite. This generalizes a theorem of Butler [20] in the theory of representations of Artinian algebras.

In this chapter $(R, \mathfrak{m}, k)$ is a Henselian CM local ring and we denote by $\mathfrak{M}$ (resp. $\mathfrak{C}$) the category of finitely generated modules (resp. CM modules) over R. We start by recalling the definition of Grothendieck groups.

(13.1) DEFINITION. Let $\mathfrak{A}$ be an additive subcategory of an Abelian category, which is skeletally small and closed under extensions. We consider an Abelian group:

$$G(\mathfrak{A}) = \oplus \mathbb{Z} \cdot X,$$

where X runs through all isomorphism classes of objects in $\mathfrak{A}$. Denote by $\mathrm{Ex}(\mathfrak{A})$ the subgroup of $G(\mathfrak{A})$ generated by

$$\{X - X' - X'' |\text{ there is an exact sequence } 0 \to X' \to X \to X'' \to 0 \text{ in } \mathfrak{A}\}.$$

The **Grothendieck group** of $\mathfrak{A}$ is defined by

$$\mathrm{K}_0(\mathfrak{A}) = G(\mathfrak{A})/\mathrm{Ex}(\mathfrak{A}).$$

We denote the class of an object X in $G(\mathfrak{A})$ by the same letter X, while the class in $\mathrm{K}_0(\mathfrak{A})$ will be denoted by $[X]$.

We are particularly concerned with the Grothendieck groups $\mathrm{K}_0(\mathfrak{M})$ and $\mathrm{K}_0(\mathfrak{C})$. By the natural embedding $i : \mathfrak{C} \to \mathfrak{M}$, one has a homomorphism of groups $\mathrm{K}_0(i) : \mathrm{K}_0(\mathfrak{C}) \to \mathrm{K}_0(\mathfrak{M})$. We remark that this is actually an isomorphism.

(13.2) LEMMA. *As a group, $K_0(\mathfrak{C})$ is isomorphic to $K_0(\mathfrak{M})$ via $K_0(i)$.*

PROOF: Let d be the dimension of R. For any module X in $\mathfrak{M}$, take a free resolution to have an exact sequence:

$$0 \to Y \to F_{d-1} \to F_{d-2} \to \cdots \to F_1 \to F_0 \to X \to 0,$$

where each F_i is free and Y is a CM module, (1.4). Then, by definition, one obtains $[X] = \sum_{i=0}^{d-1}(-1)^i[F_i] + (-1)^d[Y]$ in $K_0(\mathfrak{M})$. Thus one can define a homomorphism of groups $\varphi : G(\mathfrak{M}) \to K_0(\mathfrak{C})$ by sending X to $\sum_{i=0}^{d-1}(-1)^i[F_i] + (-1)^d[Y]$. It is an easy exercise to see that $\varphi(M)$ is independent of the choice of free resolutions and hence it induces the group homomorphism $\Phi : K_0(\mathfrak{M}) \to K_0(\mathfrak{C})$. (Check this.) Then it obviously holds that $\Phi \cdot K_0(i) = 1$ and $K_0(i) \cdot \Phi = 1$, hence $K_0(i)$ is an isomorphism. ∎

By virtue of this lemma, we usually identify the two groups $K_0(\mathfrak{M})$ and $K_0(\mathfrak{C})$. Note that, for normal domains of dimension 2, $K_0(\mathfrak{C})$ is more approachable than it is for other cases.

(13.3) LEMMA. *Let R be a normal local domain of dimension 2, and suppose the class $[k]$ of the residue field is zero in $K_0(\mathfrak{M})$. Then $K_0(\mathfrak{C})$ is isomorphic to $\mathbb{Z} \oplus \mathrm{Cl}(R)$, where $\mathrm{Cl}(R)$ denotes the divisor class group of R.*

PROOF: Recall that all modules in $\mathfrak{C}$ are reflexive, and one sees that the rank function $M \mapsto \mathrm{rank}(M)$ induces a group homomorphism $\mathrm{rk} : K_0(\mathfrak{C}) \to \mathbb{Z}$, which is surjective, since $\mathrm{rk}([R]) = 1$.

For any module X in $\mathfrak{M}$, it is known (see Matsumura [47]) that there is a finite chain of submodules:

(13.3.1) $$0 = X_0 \subset X_1 \subset \ldots \subset X_{n-1} \subset X_n = X, \quad X_i/X_{i-1} \simeq R/\mathfrak{p}_i,$$

where each $\mathfrak{p}_i$ is a prime ideal of R ($1 \leq i \leq n$). In particular, the class $[X]$ in $K_0(\mathfrak{M})$ is equal to $\sum_i [R/\mathfrak{p}_i] = \sum_i ([R] - [\mathfrak{p}_i])$. Here we may assume that all $\mathfrak{p}_i$ are of height at most one, because $[R/\mathfrak{m}] = 0$ and because R has dimension two. Consequently any element ω in $K_0(\mathfrak{M})$ can be written as

(13.3.2) $$\omega = a[R] + \sum_{\mathfrak{p}} b_{\mathfrak{p}}[\mathfrak{p}],$$

where $\mathfrak{p}$ runs through all primes of height one and a, $b_{\mathfrak{p}}$ are integers with $b_{\mathfrak{p}} = 0$ for almost all $\mathfrak{p}$, so that the summation is finite. Notice that with the notation of (13.3.2),

(13.3.3) $$\mathrm{rk}(\omega) = a + \sum_{\mathfrak{p}} b_{\mathfrak{p}}.$$

Now define a homomorphism $\varphi : K_0(\mathfrak{M}) \to \text{Cl}(R)$ as follows: For an element ω written as in (13.3.2), set $\varphi(\omega) = \sum_\mathfrak{p} b_\mathfrak{p} \cdot cl(\mathfrak{p})$, where $cl(\mathfrak{p})$ denotes the divisor class of $\mathfrak{p}$ in $\text{Cl}(R)$. It is easy to see that φ is independent of the description of ω in (13.3.2) and it actually defines a homomorphism of $K_0(\mathfrak{M})$ to $\text{Cl}(R)$. (Prove this as an exercise.) Since $\text{Cl}(R)$ is generated by the classes of primes of height one, φ is surjective. Therefore we can define a surjective homomorphism $\psi : K_0(\mathfrak{M}) \to \mathbb{Z} \oplus \text{Cl}(R)$ by $\psi(\omega) = (\text{rk}(\omega), \varphi(\omega))$.

Now we prove that ψ is injective. For this purpose, let ω be as in (13.3.2) and assume $\psi(\omega) = 0$. Since $a + \sum_\mathfrak{p} b_\mathfrak{p} = 0$, such an ω has a description

(13.3.4) $$\omega = \sum_i c_i [R/\mathfrak{p}_i] - \sum_j d_j [R/\mathfrak{q}_j],$$

where $\mathfrak{p}_i$, $\mathfrak{q}_j$ are distinct primes of height one and c_i, d_j are positive integers. Here we must have $\sum_i c_i \cdot cl(\mathfrak{p}_i) = \sum_j d_j \cdot cl(\mathfrak{q}_j)$, since $\varphi(\omega) = 0$. Setting $I = \cap_i \mathfrak{p}_i^{(c_i)}$ and $J = \cap_j \mathfrak{q}_j^{(d_j)}$ (symbolic powers), we see from the definition of divisor class groups that I is isomorphic to J as an R-module. In particular, in the group $K_0(\mathfrak{M})$, we have $[R/I] = [R/J]$. Note that, in general, for ideals A and B, $[R/A \cap B] = [R/A] + [R/B] - [R/A+B]$ in $K_0(\mathfrak{M})$, so that, if $A + B$ is $\mathfrak{m}$-primary, we have $[R/A \cap B] = [R/A] + [R/B]$ by the assumption. Applying this successively to the above, we obtain $\sum_i [R/\mathfrak{p}_i^{(c_i)}] = \sum_j [R/\mathfrak{q}_j^{(d_j)}]$ in $K_0(\mathfrak{M})$. Hence $\sum_i c_i [R/\mathfrak{p}_i] = \sum_j d_j [R/\mathfrak{q}_j]$, and ω in (13.3.4) must be zero as an element of $K_0(\mathfrak{M})$. This proves the injectivity of ψ and the lemma follows. ∎

Note that the assumption in the above lemma will be satisfied for most cases. More precisely,

(13.4) LEMMA. *With the above notation, assume that R is an excellent ring of positive dimension and the residue field k is algebraically closed. Then $[k] = 0$ in $K_0(\mathfrak{M})$.*

PROOF: Take a prime ideal $\mathfrak{p}$ of R such that $R/\mathfrak{p}$ is of dimension one. Let S be the integral closure of $R/\mathfrak{p}$ in its quotient field. Since R is excellent, S is a finite module over R. Furthermore it is a discrete valuation ring, since R, hence $R/\mathfrak{p}$, is a Henselian ring. Let t be a prime element of S. Then there is an exact sequence of S-modules:

$$0 \longrightarrow S \xrightarrow{\;t\;} S \longrightarrow S/tS \longrightarrow 0,$$

where $S/tS = k$, since it is a finite extension field of k and since k is algebraically closed. When regarding this as a sequence of R-modules, we have $[k] = [S] - [S] = 0$ in $K_0(\mathfrak{M})$ as required. ∎

(13.5) REMARK. The assumption for k in (13.4) is indispensable. Actually in the next chapter we will have an example in which $[k] \neq 0$ in $K_0(\mathfrak{M})$; see (14.12).

Now let us compute the group $K_0(\mathfrak{C})$. For this we make the following:

(13.6) DEFINITION. Define the group $AR(\mathfrak{C})$ as the subgroup of $G(\mathfrak{C})$ generated by

$$\{X - X' - X'' | \text{ there is an AR sequence } 0 \to X' \to X \to X'' \to 0 \text{ in } \mathfrak{C}\}.$$

Clearly we see that $AR(\mathfrak{C}) \subset Ex(\mathfrak{C}) \subset G(\mathfrak{C})$, while we can prove the following result.

(13.7) THEOREM. (Auslander-Reiten [10]) *If R is of finite representation type, then* $Ex(\mathfrak{C}) = AR(\mathfrak{C})$.

PROOF: Recall that there is an embedding of category $c : \mathfrak{C} \to \text{mod}(\mathfrak{C})$; see (4.6). From this, we can define a homomorphism

$$\gamma : G(\mathfrak{C}) \to K_0(\text{mod}(\mathfrak{C})),$$

by $\gamma(M) = [(\ , M)]$ for $M \in G(\mathfrak{C})$, where we use the same notation as in Chapter 4, so that $(\ , M) = \text{Hom}_R(\ , M)$. First we show that γ is injective. To do this, let M and N be in $\mathfrak{C}$ with $[(\ , M)] = [(\ , N)]$ in $K_0(\text{mod}(\mathfrak{C}))$. Then there are exact sequences in $\text{mod}(\mathfrak{C})$:

(13.7.1) $\qquad 0 \to F_i \to G_i \to H_i \to 0 \quad (1 \leq j \leq n),$

with the equality $(\ , M) + \sum_i F_i + \sum_i H_i = (\ , N) + \sum_i G_i$ in $G(\text{mod}(\mathfrak{C}))$. Substituting R into this equality, we have $M + \sum_i F_i(R) + \sum_i H_i(R) = N + \sum_i G_i(R)$ in $G(\mathfrak{C})$. Here note that the sequences of R-modules $0 \to F_i(R) \to G_i(R) \to H_i(R) \to 0$ are split. (Prove this.) Hence, $\sum_i F_i(R) + \sum_i H_i(R) = \sum_i G_i(R)$ in $G(\mathfrak{C})$. Therefore $M = N$ in $G(\mathfrak{C})$, and so γ is injective.

What are the subgroups $\gamma(Ex(\mathfrak{C}))$ and $\gamma(AR(\mathfrak{C}))$ in $K_0(\text{mod}(\mathfrak{C}))$? To see this, let

(13.7.2) $\qquad 0 \to M' \to M \to M'' \to 0$

be an exact sequence in $\mathfrak{C}$. We can define a functor F on $\mathfrak{C}$ by the exact sequence:

(13.7.3) $\qquad 0 \to (\ , M') \to (\ , M) \to (\ , M'') \to F \to 0$

It is evident that F is in $\text{mod}(\mathfrak{C})$ and $F(R) = 0$. On the other hand, we have shown in (4.9) that any object F in $\text{mod}(\mathfrak{C})$ with the property $F(R) = 0$ is obtained in this way. Therefore we see that $\gamma(\text{Ex}(\mathfrak{C}))$ is the subgroup of $\text{K}_0(\text{mod}(\mathfrak{C}))$ generated by $\{F \in \text{mod}(\mathfrak{C}) | \ F(R) = 0\}$. In the proof of (4.13) we proved that the sequence (13.7.2) is an AR sequence if and only if the functor F in (13.7.3) is a simple object in $\text{Mod}(\mathfrak{C})$. This precisely means that $\gamma(\text{AR}(\mathfrak{C}))$ is a subgroup of $\text{K}_0(\text{mod}(\mathfrak{C}))$ generated by

$$\{S \in \text{mod}(\mathfrak{C}) | \ S \text{ is a simple functor with } S(R) = 0\}.$$

Since γ is injective, to show the theorem, it is enough to prove $\gamma(\text{Ex}(\mathfrak{C})) = \gamma(\text{AR}(\mathfrak{C}))$, hence it is sufficient to show the following:

(13.7.4) For any $F \in \text{mod}(\mathfrak{C})$ with $F(R) = 0$ there is a finite series of subfunctors in $\text{mod}(\mathfrak{C})$: $0 \subset F_1 \subset F_2 \subset \ldots \subset F_{n-1} \subset F_n = F$ such that each F_i/F_{i-1} $(1 \leq i \leq n)$ is a simple object in $\text{Mod}(\mathfrak{C})$.

To prove this, consider the module $L = \oplus N$ where N runs through all the isomorphism classes of indecomposable CM modules over R. Since R is of finite representation type, L is a finitely generated module, hence it is CM. Notice that for an arbitrary $F \in \text{mod}(\mathfrak{C})$ with $F(R) = 0$, $F(L)$ is a module of finite length. In fact, if F has the presentation as in (13.7.3), then F is a subfunctor of $\text{Ext}^1_R(\ ,M')$ and thus $F(L) \subset \text{Ext}^1_R(L, M')$. Since $\text{Ext}^1_R(L, M')$ is a module of finite length by (3.3), so is $F(L)$.

We shall prove (13.7.4) by induction on the length of $F(L)$. If F is a simple functor, then there is nothing to prove. So assume that F is not simple.

Take indecomposable $M \in \mathfrak{C}$ with $F(M) \neq 0$. Then as in the proof of (4.12), one can construct an epimorphism $\pi : F \to S_M$ in $\text{Mod}(\mathfrak{C})$. Since R is of finite representation type, $\mathfrak{C}$ admits AR sequences; see the proof of (4.22). Hence, by (4.13), we have $S_M \in \text{mod}(\mathfrak{C})$. Letting $G = \text{Ker}(\pi)$, we thus see that the subfunctor G of F is in $\text{mod}(\mathfrak{C})$, (4.19). Since the length of $F(L)$ is the sum of those of $G(L)$ and $S_M(L)$, the induction hypothesis can be applied to G, thus (13.7.4) is true for G. Hence it is true as well for F, and the proof is completed. ∎

(13.8) REMARK. It has been conjectured by Auslander that the converse of the theorem is true, i.e., if $\text{Ex}(\mathfrak{C}) = \text{AR}(\mathfrak{C})$, then $\mathfrak{C}$ is of finite representation type. This was proved affirmatively for Artinian algebras by Auslander, and for complete one dimensional domains by Auslander and Reiten [10].

By the theorem, if the AR quiver Γ of R is finite, then we can compute the Grothendieck group $\text{K}_0(\mathfrak{C})$ only from the information involved in Γ.

(13.9) *Example*. Let $R = \mathbb{C}\{x, y, z\}/(x^3 + y^4 + z^2)$. Then the AR quiver for R is figured as follows, cf.(10.15).

(E_6)
$$\begin{array}{c} e \\ \updownarrow \\ d \\ \updownarrow \\ a \rightleftarrows b \rightleftarrows c \rightleftarrows f \rightleftarrows r = [R]. \end{array}$$

From this graph, there are only six AR quivers in $\mathfrak{C}$, namely

$$0 \to a \to b \to a \to 0, \qquad 0 \to b \to a \oplus c \to b \to 0,$$
$$0 \to e \to d \to e \to 0, \qquad 0 \to d \to e \oplus c \to d \to 0,$$
$$0 \to c \to b \oplus d \oplus f \to c \to 0, \qquad 0 \to f \to r \oplus c \to f \to 0.$$

Hence by Theorem (13.7), $K_0(\mathfrak{C})$ is the group generated by $\{a, b, c, d, e, f, r\}$ with the relations $2a = b$, $2b = a + c$, $2e = d$, $2d = e + c$, $2c = b + d + f$ and $2f = r + c$. Consequently, we have $K_0(\mathfrak{C}) \simeq \mathbb{Z} \oplus \mathbb{Z}/3\mathbb{Z}$, and $\mathrm{Cl}(R) \simeq \mathbb{Z}/3\mathbb{Z}$ by (13.3).

Since we already know the AR sequences for simple singularities; cf. (12.12), we can easily compute the Grothendieck groups $K_0(\mathfrak{C})$ for these rings. We end this chapter by exhibiting them below.

(13.10) PROPOSITION. $K_0(\mathfrak{C})$ *for simple singularities are shown in the following.*

The Type	Odd Dimension	Even Dimension
A_n (n :even)	$\mathbb{Z}$	$\mathbb{Z} \oplus \mathbb{Z}/(n+1)\mathbb{Z}$
A_n (n :odd)	$\mathbb{Z}^{(2)}$	$\mathbb{Z} \oplus \mathbb{Z}/(n+1)\mathbb{Z}$
D_n (n :even)	$\mathbb{Z}^{(3)}$	$\mathbb{Z} \oplus (\mathbb{Z}/2\mathbb{Z})^{(2)}$
D_n (n :odd)	$\mathbb{Z}^{(2)}$	$\mathbb{Z} \oplus \mathbb{Z}/4\mathbb{Z}$
E_6	$\mathbb{Z}$	$\mathbb{Z} \oplus \mathbb{Z}/3\mathbb{Z}$
E_7	$\mathbb{Z}^{(2)}$	$\mathbb{Z} \oplus \mathbb{Z}/2\mathbb{Z}$
E_8	$\mathbb{Z}$	$\mathbb{Z}$

Chapter 14. CM modules on quadrics

In the case when R is a hypersurface defined by a quadratic form, CM R-modules are rather easy to handle. In fact, we shall show in this chapter that they correspond to Clifford modules. All ideas below are taken from Buchweitz-Eisenbud-Herzog [18].

Throughout the chapter k denotes an *arbitrary field of characteristic unequal to* 2 and (V, Q) is a quadratic space over k, i.e., V is a k-vector space of finite dimension and Q is a mapping $V \to k$ with the properties:

(14.1.1) $Q(ax) = a^2 Q(x)$ $(a \in k, x \in V)$, and

(14.1.2) $(x, y)_Q := \frac{1}{2}\{Q(x+y) - Q(x) - Q(y)\}$ is a symmetric bilinear form on V.

Note that if V has a basis $\{e_1, e_2, \ldots, e_n\}$ and if it has the dual basis $\{x_1, x_2, \ldots, x_n\}$, then Q is a symmetric form:

$$\sum_{i,j=1}^{n} a_{ij} x_i x_j \quad (a_{ij} \in k,\ a_{ij} = a_{ji} \text{ for any } i, j).$$

Hence after a change of basis, we may write Q as

(14.1.3) $$\sum_{i=1}^{n} \alpha_i x_i^2 \quad (\alpha_i \in k).$$

We are interested in the rings defined by quadratic forms.

(14.1) DEFINITION. Let (V, Q) and $\{x_1, x_2, \ldots, x_n\}$ be as above. Then we define:

(14.1.4) $S = k[[x_1, x_2, \ldots, x_n]]$ (the formal power series ring),

(14.1.5) $R_Q = S/QS$.

As before we denote the category of CM modules over R_Q by $\mathfrak{C}(R_Q)$. We are concerned with the representation type of this category. So, first of all, we need the condition for R_Q to be an isolated singularity; cf. (4.22).

(14.2) LEMMA. *R_Q has only an isolated singularity if and only if the bilinear form $(\ ,\ )_Q$ is nondegenerate.*

PROOF: Let Q have the description as in (14.1.3). Then R_Q is an isolated singularity if and only if the ideal of R_Q generated by $\{\partial Q/\partial x_i | \ 1 \leq i \leq n\} = \{2\alpha_i x_i | \ 1 \leq i \leq n\}$ is primary belonging to the maximal ideal. This is equivalent to $\alpha_i \neq 0$ for all i, which occurs only when $(\ ,\ )_Q$ is nondegenerate. ∎

Recall that CM modules over R_Q without free summands are obtained by reduced matrix factorizations of Q, see (7.6). We show below that these matrix factorizations can be chosen, up to equivalence, so that all their entries are linear forms. For example, let $S = k[[x, y]]$ and $Q = x^2 + y^2$. Then a pair of matrices

$$\left(\begin{pmatrix} x + xy & -y - x^2 y \\ y & x - xy \end{pmatrix}, \begin{pmatrix} x - xy & y + x^2 y \\ -y & x + xy \end{pmatrix} \right)$$

is certainly a matrix factorization of Q. However this is equivalent to

$$\left(\begin{pmatrix} x & -y \\ y & x \end{pmatrix}, \begin{pmatrix} x & y \\ -y & x \end{pmatrix} \right),$$

which has linear entries.

(14.3) PROPOSITION. (Buchweitz-Eisenbud-Herzog [18]) *With the notation in (14.1), suppose that R_Q is an isolated singularity. Then every reduced matrix factorization of Q is equivalent to one with linear entries.*

Before proving the proposition, we remark that CM modules with 'linear matrix factorization' have a specific property. For this, let us denote by G_Q the associated graded ring of R_Q along the maximal ideal, so that $G_Q = k[x_1, x_2, \ldots, x_n]/(Q)$. Note that R_Q is an isolated singularity if and only if G_Q is.

(14.4) LEMMA. *Let (φ, ψ), (φ', ψ') be reduced matrix factorizations of Q whose entries are linear forms and let M, M' be the CM modules corresponding to them: $M = \mathrm{Coker}(\varphi, \psi)$, $M' = \mathrm{Coker}(\varphi', \psi')$. Consider the graded modules*

$$N = \mathrm{Coker}(G_Q(-1)^{(n)} \xrightarrow{\varphi} G_Q^{(n)}),$$

$$N' = \mathrm{Coker}(G_Q(-1)^{(n')} \xrightarrow{\varphi'} G_Q^{(n')}),$$

over the graded ring G_Q. Then:

(14.4.1) *M is isomorphic to the completion $\widehat{N}$ of N with respect to the irrelevant maximal ideal, and*

$$\text{Ext}^i_{G_Q}(N, N')\widehat{} \simeq \text{Ext}^i_{R_Q}(M, M')$$

for any i.

(14.4.2) *If R_Q is an isolated singularity, then $\text{Ext}^i_{G_Q}(N, N')$ is an Artinian G_Q-module for any $i > 0$.*

(14.4.3) *There is a graded G_Q-free resolution of N of the form:*

$$\cdots \longrightarrow G_Q(-3)^{(n)} \xrightarrow{\varphi} G_Q(-2)^{(n)} \xrightarrow{\psi} G_Q(-1)^{(n)} \xrightarrow{\varphi} G_Q^{(n)} \longrightarrow N \longrightarrow 0.$$

(14.4.4) *There is an isomorphism of graded G_Q-modules:*

$$\text{Ext}^i_{G_Q}(N, N') \simeq \text{Ext}^{i+2}_{G_Q}(N, N')(-2),$$

for each $i > 0$.

PROOF: For simplicity write G instead of G_Q. Since $G = k[x_1, x_2, \ldots, x_n]/(Q)$, we have $\widehat{G} = R_Q$. Hence (14.4.1) is evident from this.

In $\text{Ext}^i_G(N, N')\widehat{} \simeq \text{Ext}^i_{R_Q}(M, M')$, the latter module is of finite length if $i > 0$, since R_Q is an isolated singularity; see (3.3). In general, a graded G-module is Artinian if and only if its completion is Artinian. Hence (14.4.2) follows.

We leave the proof of (14.4.3) to the reader, since it is proved by the completely same argument as in (7.2).

By (14.4.3) we have an exact sequence of graded G-modules:

$$0 \longrightarrow N(-2) \longrightarrow G(-1)^{(n)} \xrightarrow{\varphi} G^{(n)} \longrightarrow N \longrightarrow 0.$$

Hence it follows that $\text{Ext}^i_G(N(-2), N') \simeq \text{Ext}^{i+2}_G(N, N')$ for any $i > 0$, which shows (14.4.4). ∎

(14.5) LEMMA. *Let N, N' be as in (14.4).*

(14.5.1) *If $\text{Ext}^i_{G_Q}(N, N')$ ($i > 0$) are Artinian G_Q-modules, then, for any $i > 0$,*

$$\text{Ext}^i_{G_Q}(N, N')_j = 0 \quad \text{whenever} \quad i + j > 0.$$

(14.5.2) *If R_Q is an isolated singularity, then $\underline{\text{Hom}}_{G_Q}(N, N')_j = 0$ for $j > 0$. (See (3.7) for the definition of $\underline{\text{Hom}}$.)*

PROOF: (14.5.1) : We use induction on n. If $n = 1$, then $Q = \alpha_1 x_1^2$ ($\alpha_1 \in k$), hence every matrix factorization of Q is a direct sum of matrix factorizations of the form

$(\beta x_1, \gamma x_1)$ $(\beta, \gamma \in k, \alpha_1 = \beta \cdot \gamma)$. If $\alpha_1 = 0$, then N, N' are free modules and the claim is obvious. So suppose $\alpha_1 \neq 0$, and thus we may assume that $N = N' = G/x_1G$. Then it is easy to see that $\operatorname{Ext}_G^i(N, N') \simeq k(i)$ (the module k shifted degree by i), which proves (14.5.1) for $n = 1$.

Assume $n \geq 2$. Note form (14.4.4) that it is sufficient to prove (14.5.1) for $i \geq 2$. One can choose a linear form z in G that is a nonzero divisor on G, hence on N and N'. (Why?) Set $\overline{G} = G/zG$, $\overline{N} = N/zN$ and $\overline{N'} = N'/zN'$. From the short exact sequence of graded G-modules:

$$0 \longrightarrow N(-1) \xrightarrow{z} N \longrightarrow \overline{N} \longrightarrow 0,$$

we have a long one:

(*) $\quad \cdots \to \operatorname{Ext}_G^{i-1}(N, N')(1) \to \operatorname{Ext}_G^i(\overline{N}, N') \to \operatorname{Ext}_G^i(N, N') \xrightarrow{z} \operatorname{Ext}_G^i(N, N')(1) \to \cdots$.

Here an easy computation shows that

(**) $\qquad \operatorname{Ext}_G^i(\overline{N}, N') \simeq \operatorname{Ext}_{\overline{G}}^{i-1}(\overline{N}, \overline{N'})(1) \quad \text{for any } i \geq 0.$

(Use the fact that $\operatorname{Ext}_G^j(\overline{G}, N') \simeq \overline{N'}(1)$ $(j = 1)$ and $\simeq 0$ (otherwise) to the spectral sequence:

$$\operatorname{Ext}_{\overline{G}}^p(\overline{N}, \operatorname{Ext}_G^q(\overline{G}, N')) \Rightarrow \operatorname{Ext}_G^{p+q}(\overline{N}, N').)$$

Since $\operatorname{Ext}_G^i(N, N')$ $(i > 0)$ are Artinian modules, we see from (*) and from (**) that $\operatorname{Ext}_{\overline{G}}^i(\overline{N}, \overline{N'})$ $(i > 0)$ are also Artinian. Thus we may apply the induction hypothesis to the $\overline{G}$-modules $\overline{N}$ and $\overline{N'}$ to get:

$$\operatorname{Ext}_{\overline{G}}^i(\overline{N}, \overline{N'})_j = 0 \quad \text{whenever } i + j > 0 \text{ and } i > 0.$$

Therefore, letting $K^i = \operatorname{Ker}(\operatorname{Ext}_G^i(N, N') \xrightarrow{z} \operatorname{Ext}_G^i(N, N')(1))$, we see from (*) that

$$(K^i)_j = 0 \quad \text{if} \quad i + j > 0 \text{ and } i \geq 2.$$

Let ξ be an arbitrary element in $\operatorname{Ext}_G^i(N, N')_j$ with $i + j > 0$ and with $i \geq 2$. Since $\operatorname{Ext}_G^i(N, N')$ is an Artinian G-module, a power of z kills ξ, i.e. $z^l \xi = 0$ for some $l > 0$. Then, since $z^{l-1} \xi \in (K^i)_{j+l-1}$, we have $z^{l-1} \xi = 0$ by the above. Continueing this, we can show that $\xi = 0$, hence (14.5.1) is proved.

(14.5.2) : Let $L' = \text{Coker}(G(-1)^{(n')} \xrightarrow{\psi'} G^{(n')})$. Then, by (14.4.3), we have the exact sequence of graded G-modules:

$$0 \longrightarrow L'(-1) \longrightarrow G^{(n')} \longrightarrow N' \longrightarrow 0,$$

therefore the following is also exact:

$$\text{Hom}_G(N, G^{(n')}) \longrightarrow \text{Hom}_G(N, N') \longrightarrow \text{Ext}^1_G(N, L')(-1).$$

Then it follows from (4.15.1) that $\underline{\text{Hom}}_G(N, N')$ is a submodule of $\text{Ext}^1_G(N, L')(-1)$. Thus to prove (14.5.2) it is enough to show that

$$\text{Ext}^1_G(N, L')(-1)_j = 0 \quad \text{if} \quad j > 0.$$

However this is a direct consequence of (14.4.2) and (14.5.1). ∎

We are now ready to prove the proposition.

PROOF OF PROPOSITION (14.3): Let (φ, ψ) be a reduced matrix factorization of Q of size n. Write

$$\varphi = \sum_{i=1}^{\infty} \varphi_i, \qquad \psi = \sum_{i=1}^{\infty} \psi_i,$$

where φ_i, ψ_i are matrices of forms of degree i. Taking the homogeneous part of the equation $\varphi \cdot \psi = \psi \cdot \varphi = QI$, we see that

$$\varphi_1 \cdot \psi_1 = \psi_1 \cdot \varphi_1 = QI,$$

so that (φ_1, ψ_1) is also a matrix factorization of Q. Let M be the CM module $\text{Coker}(\varphi_1, \psi_1)$ and set $N = \text{Coker}(G_Q(-1)^{(n)} \xrightarrow{\varphi_1} G_Q^{(n)})$ as in (14.4). Notice that by (14.4.3), there is an exact sequences of G_Q-modules:

$$(*) \qquad \cdots \xrightarrow{\varphi_1} G_Q(-2)^{(n)} \xrightarrow{\psi_1} G_Q(-1)^{(n)} \xrightarrow{\varphi_1} G_Q^{(n)} \longrightarrow N \longrightarrow 0.$$

We will show that (φ, ψ) is equivalent to (φ_1, ψ_1).

Looking at the degree 3 part in $\varphi \cdot \psi = \psi \cdot \varphi = QI$, we have $\varphi_1 \cdot \psi_2 + \varphi_2 \cdot \psi_1 = 0$ and $\psi_1 \cdot \varphi_2 + \psi_2 \cdot \varphi_1 = 0$ which means the pair of matrices $(\varphi_2, -\psi_2)$ gives a chain map from the complex (*) to itself:

$$\begin{array}{ccccccccc}
\cdots \xrightarrow{\psi_1} & G_Q(-3)^{(n)} & \xrightarrow{\varphi_1} & G_Q(-2)^{(n)} & \xrightarrow{\psi_1} & G_Q(-1)^{(n)} & \xrightarrow{\varphi_1} & G_Q^{(n)} \\
& \varphi_2 \downarrow & & -\psi_2 \downarrow & & \varphi_2 \downarrow & & \\
\cdots \xrightarrow{\varphi_1} & G_Q(-1)^{(n)} & \xrightarrow{\psi_1} & G_Q^{(n)} & \xrightarrow{\varphi_1} & G_Q(1)^{(n)}. & &
\end{array}$$

Therefore it gives an element of degree 2 in $\mathrm{Ext}^1_{G_Q}(N,N)$, which must be zero by (14.5.1); equivalently the chain map is homotopically zero. Thus there are square matrices α_2, β_2 consisting of forms in S such that the following equations hold as matrices on G_Q:

$$\varphi_2 = \alpha_2 \cdot \varphi_1 - \varphi_1 \cdot \beta_2, \qquad -\psi_2 = -\beta_2 \cdot \psi_1 + \psi_1 \cdot \alpha_2.$$

Then these hold true as matrices on R_Q, because $\hat{G}_Q = R_Q$. Now set $\varphi^{(2)} = (1-\alpha_2)\varphi(1-\beta_2)^{-1}$ and $\psi^{(2)} = (1-\beta_2)\psi(1-\alpha_2)^{-1}$. They are well-defined matrices on S. In fact, $1-\alpha_2, 1-\beta_2$ have inverses $\sum_{i=0}^\infty \alpha_2^i, \sum_{i=0}^\infty \beta_2^i$ respectively as matrices on S. (S is complete !) Note that $(\varphi^{(2)}, \psi^{(2)})$ is again a matrix factorization of Q without homogeneous part of degree 2, so that the matrices are written:

$$\varphi^{(2)} = \varphi_1 + \sum_{i=3}^\infty \varphi_i^{(2)}, \qquad \psi^{(2)} = \psi_1 + \sum_{i=3}^\infty \psi_i^{(2)},$$

where $\varphi_i^{(2)}, \psi_i^{(2)}$ are matrices of forms of degree i.

This procedure can be continued inductively to get:

(14.3.1) For any $j \geq 1$, there is a matrix factorization $(\varphi^{(j)}, \psi^{(j)})$ of Q (with $(\varphi^{(1)}, \psi^{(1)}) = (\varphi, \psi)$) which has the description:

$$\varphi^{(j)} = \varphi_1 + \sum_{i=j+1}^\infty \varphi_i^{(j)}, \qquad \psi^{(j)} = \psi_1 + \sum_{i=j+1}^\infty \psi_i^{(j)},$$

where $\varphi_i^{(j)}, \psi_i^{(j)}$ are matrices consisting of forms of degree i. Furthermore there is a pair of matrices (α_j, β_j) with $\varphi^{(j)} = (1-\alpha_j)\varphi^{(j-1)}(1-\beta_j)^{-1}$ and $\psi^{(j)} = (1-\beta_j)\psi^{(j-1)}(1-\alpha_j)^{-1}$. (Give a proof of this .)

Since S is complete, we can take

$$A = \lim_{n\to\infty} \prod_{j=2}^n (1-\alpha_j), \qquad B = \lim_{n\to\infty} \prod_{j=2}^n (1-\beta_j),$$

Then they are invertible matrices on S and satisfy

$$A\varphi B^{-1} = \lim_{n\to\infty} \varphi^{(j)} = \varphi_1 \quad \text{and} \quad B\psi A^{-1} = \lim_{n\to\infty} \psi^{(j)} = \psi_1$$

Therefore the matrix factorization (φ, ψ) is equivalent to (φ_1, ψ_1) as required. ∎

Assume that R_Q has only an isolated singularity and suppose we are given a CM module M over R_Q with no free summands. Then the corresponding matrix factorization (φ, ψ) can be chosen to be reduced. By virtue of (14.3) we may assume that the entries φ_{ij}, ψ_{ij} are all linear forms, hence are elements in the dual space V^* of V. Let m be the size of the matrices φ, ψ and prepare the k-vector space U of dimension m with some fixed basis. Then the above says that φ, ψ give linear maps:

(14.6.1) $\qquad V \longrightarrow \mathrm{End}_k(U) \; ; \; v \mapsto \varphi(v) = (\varphi_{ij}(v)), \; \psi(v) = (\psi_{ij}(v)).$

Let $k\langle V \rangle$ denote the free k-algebra on V, let W_0, W_1 be copies of U and set $W = W_0 \oplus W_1$. We define a $k\langle V \rangle$-module structure on W as follows:

(14.6.2) An element v in V acts on W_0 by $\varphi(v) : W_0 \to W_1$ and acts on W_1 by $\psi(v) : W_1 \to W_0$.

Notice that $\psi(v)\varphi(v) = Q(v)1_{W_0}$ and $\varphi(v)\psi(v) = Q(v)1_{W_1}$ for any v in V. Hence,

(14.6.3) $v^2 - Q(v)$ acts trivially on W for any $v \in V$.

In other words, W is a module over the ring

(14.6.4) $\qquad\qquad\qquad k\langle V \rangle / (v^2 - Q(v)| \, v \in V).$

(14.6) DEFINITION. The k-algebra defined by (14.6.4) is called the **Clifford algebra** over the quadratic space (V, Q), and is denoted by $C(V, Q)$ or $C(Q)$. Define the degree of an element $v_1 v_2 \cdots v_n \in C(Q)$ ($v_i \in V$, $1 \le i \le n$) as n (mod 2), which is well-defined by (14.6.4). Write $C_0(Q)$ (resp. $C_1(Q)$) as the subspace of $C(Q)$ generated by all homogeneous elements in $C(Q)$ of degree 0 (resp. 1). It is easily seen that $C_0(Q)$ is a subalgebra of $C(Q)$ and that $C_1(Q)$ is a $C_0(Q)$-module. Furthermore $C(Q) = C_0(Q) \oplus C_1(Q)$ is a $\mathbb{Z}/2\mathbb{Z}$-graded algebra over k.

We denote by $\Delta(\varphi, \psi)$ the $C(Q)$-module W defined by (14.6.2). It is clear from the definition that $\Delta(\varphi, \psi)$ is a $\mathbb{Z}/2\mathbb{Z}$-graded $C(Q)$-module by defining graded pieces: $\Delta(\varphi, \psi)_0 = W_0$, $\Delta(\varphi, \psi)_1 = W_1$.

Next let $(\alpha, \beta) : (\varphi, \psi) \to (\varphi', \psi')$ be a morphism between reduced matrix factorizations with linear entries:

$$\begin{array}{ccccc} S^{(m)} & \xrightarrow{\psi} & S^{(m)} & \xrightarrow{\varphi} & S^{(m)} \\ {\scriptstyle \alpha}\downarrow & & {\scriptstyle \beta}\downarrow & & {\scriptstyle \alpha}\downarrow \\ S^{(m')} & \xrightarrow{\psi'} & S^{(m')} & \xrightarrow{\varphi'} & S^{(m')}. \end{array}$$

Let α_0, β_0 be the constant terms of the matrices α, β respectively, (that is, α_0 is the matrix obtained from α by substituting 0 in all variables x_i in α). Then define a k-linear map $\Delta(\alpha,\beta) : \Delta(\varphi,\psi) \to \Delta(\varphi',\psi')$ by $\Delta(\alpha,\beta)(w_0) = \beta_0(w_0) \in \Delta(\varphi',\psi')_0$ ($w_0 \in \Delta(\varphi,\psi)_0$) and $\Delta(\alpha,\beta)(w_1) = \alpha_0(w_1) \in \Delta(\varphi',\psi')_1$ ($w_1 \in \Delta(\varphi,\psi)_1$). Since $\alpha_0 \cdot \varphi(v) = \varphi'(v) \cdot \beta_0$ and $\beta_0 \cdot \psi(v) = \psi'(v) \cdot \alpha_0$ for any $v \in V$, we easily see that $\Delta(\alpha,\beta)(vw) = v\Delta(\alpha,\beta)(w)$ ($v \in V, w \in \Delta(\varphi,\psi)$), hence $\Delta(\alpha,\beta)$ is a $C(Q)$-homomorphism which is obviously $\mathbb{Z}/2\mathbb{Z}$-graded.

Let $\mathfrak{grM}(C(Q))$ denotes the category of finitely generated $\mathbb{Z}/2\mathbb{Z}$-graded modules and degree-preserving homomorphisms over $C(Q)$. Recall that $\underline{\mathrm{RMF}}_S(Q)$ is the category of reduced matrix factorizations; see (7.3.2). Notice that we showed in (7.4) that $\underline{\mathrm{RMF}}_S(Q)$ is equivalent to the category $\underline{\mathfrak{C}}(R)$. Under the assumption that R_Q is an isolated singularity, we have defined above the functor from $\underline{\mathrm{RMF}}_S(Q)$ into $\mathfrak{grM}(C(Q))$:

(14.7.1) $$\Delta : \underline{\mathrm{RMF}}_S(Q) \longrightarrow \mathfrak{grM}(C(Q)).$$

(14.7) THEOREM. (Buchweitz-Eisenbud-Herzog [18]) *Suppose R_Q is an isolated singularity. Then the above functor gives rise to an equivalence of categories.*

PROOF: We construct a functor in the reverse direction. Let $W = W_0 \oplus W_1$ be an object in $\mathfrak{grM}(C(Q))$. Since it is graded, any v in V determines k-linear maps $\varphi(v) : W_0 \to W_1$ and $\psi(v) : W_1 \to W_0$. By the definition of $C(Q)$, one sees that $\varphi(v) \cdot \psi(v)$, $\psi(v) \cdot \varphi(v)$ are the multiplication maps by $Q(v)$ on W_1, W_0 respectively. Taking $v \in V$ as $Q(v) \neq 0$, we see that W_0 and W_1 are isomorphic as k-vector spaces and we may identify them: $U = W_0 = W_1$. Since $\varphi(v)$, $\psi(v)$ are linear in v, fixing a base of U, we may write $\varphi = (\varphi_{ij})$, $\psi = (\psi_{ij})$, where φ_{ij}, ψ_{ij} are all linear functions on V, hence are linear forms in S. Thus (φ,ψ) gives a reduced matrix factorization of Q with linear entries. We will denote this by $\Theta(W)$.

For a graded $C(Q)$-homomorphism $f : W = W_0 \oplus W_1 \to W' = W'_0 \oplus W'_1$ we define a morphism $\Theta(f) : \Theta(W) \to \Theta(W')$ as follows: Since f preserves degree, it gives k-linear maps $f_0 : W_0 \to W'_0$ and $f_1 : W_1 \to W'_1$. They satisfy $f_1(vw_0) = vf_0(w_0)$, $f_0(vw_1) = vf_1(w_1)$ ($v \in V$, $w_0 \in W_0$, $w_1 \in W_1$), since f is $C(Q)$-linear. Therefore $f_1 \cdot \varphi(v) = \varphi'(v) \cdot f_0$, $f_0 \cdot \psi(v) = \psi'(v) \cdot f_1$ as mappings on W_0, W_1 respectively. Regarding f_0, f_1 as matrices with constant entries in S, we see from this that (f_0, f_1) is a morphism of matrix factorizations. Write this as $\Theta(f)$. Thus we have defined a functor $\Theta : \mathfrak{grM}(C(Q)) \to \underline{\mathrm{RMF}}_S(Q)$.

By the construction it is easy to see that $\Delta(\Theta(W)) = W$ and $\Delta(\Theta(f)) = f$ for a graded $C(Q)$-module W and a graded $C(Q)$-homomorphism f, so that $\Delta \cdot \Theta$ is the identity functor on $\mathfrak{grM}(C(Q))$. It is also obvious that $\Theta\Delta((\varphi,\psi)) = (\varphi,\psi)$ for a matrix factorization (φ,ψ) with linear entries. Hence, to prove the theorem, it remains to show that $\Theta\Delta((\alpha,\beta)) = (\alpha,\beta)$ for a morphism $(\alpha,\beta) : (\varphi,\psi) \to (\varphi',\psi')$ of linear matrix factorizations. Let α_0, β_0 be the constant terms of α, β as in the definition of Δ. Then by the construction of Θ, it can be seen that $\Theta\Delta((\alpha,\beta)) = (\alpha_0, \beta_0)$. We have to show the equality $(\alpha, \beta) = (\alpha_0, \beta_0)$ as a morphism in $\underline{\mathrm{RMF}}_S(Q)$. For this, let $M = \mathrm{Coker}(\varphi, \psi)$, $M' = \mathrm{Coker}(\psi', \varphi')$ and let

$$N = \mathrm{Coker}(G_Q(-1)^{(n)} \xrightarrow{\varphi} G_Q^{(n)}),$$

$$N' = \mathrm{Coker}(G_Q(-1)^{(n')} \xrightarrow{\varphi'} G_Q^{(n')})$$

as in (14.4). Recall that the functor $\mathrm{Coker} : \underline{\mathrm{RMF}}_S(Q) \to \underline{\mathfrak{C}}(R_Q)$ is an equivalence of categories, (7.4). Hence it is sufficient to show that $\mathrm{Coker}(\alpha, \beta) = \mathrm{Coker}(\alpha_0, \beta_0)$ as an element in $\underline{\mathrm{Hom}}_{R_Q}(M, M')$. Let us write α, β as

$$\alpha = \sum_{i=0}^{\infty} \alpha_i, \qquad \beta = \sum_{i=0}^{\infty} \beta_i$$

where α_i, β_i are matrices consisting of forms of degree i. Looking at the degree $i+1$ part in $\alpha \cdot \varphi = \varphi' \cdot \beta$, $\beta \cdot \psi = \psi' \cdot \alpha$, we see that (α_i, β_i) is a morphism of matrix factorizations for any i. It is thus easy to see that (α_i, β_i) gives an element in $\underline{\mathrm{Hom}}_{G_Q}(N, N')$ of degree i, and through the completion map $\underline{\mathrm{Hom}}_{G_Q}(N, N') \to \underline{\mathrm{Hom}}_{R_Q}(M, M')$ it goes to $\mathrm{Coker}(\alpha_i, \beta_i)$. However we know from (14.5.2) that $\underline{\mathrm{Hom}}_{G_Q}(N, N')_i = 0$ $(i > 0)$, hence $\mathrm{Coker}(\alpha_i, \beta_i) = 0$ $(i > 0)$. Therefore we conclude that $\mathrm{Coker}(\alpha, \beta) = \mathrm{Coker}(\alpha_0, \beta_0)$ as desired. ∎

Recall that if $(\ ,\ )_Q$ is nondegenerate, then the Clifford algebra $C(Q)$ is a graded central simple algebra over k, so that the objects in $\mathfrak{gr}\mathfrak{M}(C(Q))$ are completely reducible and $\mathfrak{gr}\mathfrak{M}(C(Q))$ has only a finite number of simple objects; see Lam [46]. Hence we see from the theorem and from (14.2) that *the CM ring R_Q is of finite representation type if and only if the bilinear form $(\ ,\ )_Q$ is nondegenerate*.

In the rest of this chapter we assume that all quadratic forms have nondegenerate bilinear forms. If the quadratic form (V, Q) is isomorphic to the direct sum of (V', Q') with the hyperbolic space H, then the Clifford algebra $C(Q)$ is known to be isomorphic to $C(Q') \bar{\otimes}_k C(H)$ as a graded k-algebra, where $\bar{\otimes}$ denotes the graded tensor of graded algebras; see Lam [46]. Here $C(H)$ is the total matrix algebra, hence the category $\mathfrak{gr}\mathfrak{M}(C(Q))$ is Morita equivalent to $\mathfrak{gr}\mathfrak{M}(C(Q'))$. Recalling that the Witt ring $W(k)$ of k is defined to be the set of all isomorphism classes of quadratic spaces over k modulo the hyperbolic spaces, we have shown:

(14.8) PROPOSITION. *Suppose that quadratic forms (V, Q) and (V', Q') define the same class in the Witt ring $W(k)$. Then the categories $\underline{\mathrm{RMF}}_S(Q)$ and $\underline{\mathrm{RMF}}_{S'}(Q')$ are equivalent.*

(14.9) REMARK. By the proposition $W(k)$ classifies the categories of nonfree CM modules over the rings defined by quadratic forms. However one can show that they can be classified by a smaller group than $W(k)$. To see this, recall that the Brauer-Wall group $BW(k)$ is the group consisting of all classes of graded central simple algebras over k modulo graded Morita equivalence. Note that, by definition, there is a natural epimorphism of groups $W(k) \to BW(k)$ which sends the class of (V,Q) to the class of $C(Q)$. For example, $W(k) \simeq BW(k) \simeq \mathbb{Z}/2\mathbb{Z}$ when k is an algebraically closed field, and $W(\mathbb{R}) \simeq \mathbb{Z}$, $BW(\mathbb{R}) \simeq \mathbb{Z}/8\mathbb{Z}$; see Lam [46]. If the Clifford algebras $C(Q)$ and $C(Q')$ have the same class in $BW(k)$, then it is obvious by (14.7) that $\underline{\mathrm{RMF}}_S(Q)$ is equivalent to $\underline{\mathrm{RMF}}_{S'}(Q')$.

Recall that $n(R)$ denotes the number of classes of indecomposable CM modules over R. For the rings defined by quadratic forms we can easily evaluate these numbers.

(14.10) PROPOSITION. *Suppose the ring R_Q has only an isolated singularity. Consider the condition:*

(∗) *either $\dim_k(V)$ is odd, or $\dim_k(V)$ is even with $(-1)^{\dim(V)/2} \det(Q) \notin (k^*)^2$*

Then $n(R_Q) = 2$ under the condition (∗), but otherwise $n(R_Q) = 3$.

PROOF: Note from (14.7) that $n(R_Q)$ is bigger than the number of classes of simple objects in $\mathfrak{gr}\mathfrak{M}(C(Q))$, by just one. Since $C(Q)$ is a graded Azumaya algebra, there is only one class of simple module over $C(Q)$ up to degree shifting. Therefore $\mathfrak{gr}\mathfrak{M}(C(Q))$ has at most two simple objects. More precisely, if $W = W_0 \oplus W_1$ is a simple object in $\mathfrak{gr}\mathfrak{M}(C(Q))$, then shifting the degree, the module $\tilde{W} = W_1 \oplus W_0$ is also simple in $\mathfrak{gr}\mathfrak{M}(C(Q))$. Hence $W, \tilde{W}$ are all of the simple modules. If $W \simeq \tilde{W}$ as a $C(Q)$-module then $n(R_Q) = 2$, and otherwise $n(R_Q) = 3$.

It is known from the theory of quadratic forms [46] that the even Clifford algebra $C_0(Q)$ is an Azumaya k-algebra when (∗) is satisfied, otherwise $C_0(Q)$ is a product of two Azumaya algebras. Thus, assuming the condition (∗), we see that $W_0 \simeq W_1$ as a $C_0(Q)$-module, hence $W \simeq W_0 \otimes_{C_0(Q)} C(Q) \simeq \tilde{W}$ and consequently $n(R_Q) = 2$. When (∗) does not hold, if $C_0(Q) \simeq C' \times C'''$ with C', C''' Azumaya algebras over k, then there are nonisomorphic $C_0(Q)$-modules W', W'', hence $W' \otimes_{C_0(Q)} C(Q)$ and $W'' \otimes_{C_0(Q)} C(Q)$ are nonisomorphic simple objects. Therefore $n(R_Q) = 3$. ∎

(14.11) *Example.* If k is an algebraically closed field, then every quadratic form is equivalent to $\sum_{i=1}^m x_i^2 \in k[x_1, x_2, \ldots, x_n]$ ($m \leq n$). Thus if R_Q is an isolated singularity, then $R_Q = k[[x_1, x_2, \ldots, x_n]]/(x_1^2 + x_2^2 + \ldots + x_n^2)$. By the above proposition we have $n(R_Q) = 2$ if n is odd, and $n(R_Q) = 3$ if n is even. This is a special case of the Knörrer's periodicity (12.10).

(14.12) *Example.* Consider the case when k is the field of real numbers **R**. Then each quadratic form is equivalent to

$$Q = x_1^2 + x_2^2 + \ldots + x_p^2 - x_{p+1}^2 - x_{p+2}^2 - \ldots - x_{p+q}^2 \in \mathbf{R}[x_1, x_2, \ldots, x_n].$$

Note that R_Q is an isolated singularity if and only if $n = p+q$. By (14.10) one can verify that $n(R_Q) = 3$ if $p - q$ is a multiple of 4, otherwise $n(R_Q) = 2$.

Furthermore one can easily describe the AR quivers for several rings of this type. For example, if $p = 2$ and $q = 0$ so that $R = R_Q = \mathbf{R}[[x, y]]/(x^2 + y^2)$, then the AR quiver of R is shown in (14.12.1), where $\mathfrak{m}$ is the maximal ideal of R. It is easy to see that $\mathrm{End}_R(\mathfrak{m})/\mathrm{rad}(\mathrm{End}_R(\mathfrak{m})) \simeq \mathbf{C}$. The indices in the diagram are attached as defined in (5.3). Because it follows from this diagram that $0 \to \mathfrak{m} \to R^2 \to \mathfrak{m} \to 0$ is the unique AR sequence in $\mathfrak{C}(R)$, we can compute the Grothendieck group of $\mathfrak{C}(R)$ by the method developed in (13.7). Actually $K_0(\mathfrak{C}(R))$ is an Abelian group generated by $[R]$ and $[\mathfrak{m}]$ with relation $2[R] = 2[\mathfrak{m}]$. It, then, follows that the class of the residue field of R is nonzero in $K_0(\mathfrak{C}(R))$, since $[R/\mathfrak{m}] = [R] - [\mathfrak{m}] \neq 0$. This gives the example announced in (13.5).

$$R_Q \underset{(1,2)}{\overset{(2,1)}{\rightleftarrows}} \mathfrak{m} \circlearrowright$$

(14.12.1)

Chapter 15. Graded CM modules on graded CM rings

We are concerned in this chapter with graded CM modules over a graded CM ring and study their relationship with completed CM modules. If a graded ring is of finite representation type, then we can show that every complete CM module comes from graded one; see (15.14).

Let $R = \sum_{i=0}^{\infty} R_i$ be an arbitrary N-graded CM ring with $R_0 = k$ a field. Setting $\mathfrak{m} = \sum_{i>0} R_i$, we denote by $\hat{R}$ the completion of R in $\mathfrak{m}$-adic topology. As before, let $\mathfrak{C}(\hat{R})$ be the category of CM $\hat{R}$-modules and $\hat{R}$-homomorphisms, and let $\mathfrak{grC}(R)$ denote the category of graded CM modules over R and graded homomorphisms preserving degree. Likewise $\mathfrak{M}(\hat{R})$ (resp. $\mathfrak{grM}(R)$) is the category of finitely generated $\hat{R}$-modules (resp. finitely generated graded R-modules) and $\hat{R}$-homomorphisms (resp. graded R-homomorphisms preserving degree). In addition, for $M \in \mathfrak{grM}(R)$ and $n \in \mathbb{Z}$, $M(n) \in \mathfrak{grM}(R)$ is defined by $M(n)_i = M_{n+i}$. Graded modules $M, N \in \mathfrak{grM}(R)$ (resp. $\in \mathfrak{grC}(R)$) are said to be **isomorphic up to degree shifting** if $M(n) \simeq N$ in $\mathfrak{grM}(R)$ (resp. in $\mathfrak{grC}(R)$) for some integer n. Notice that, for $M, N \in \mathfrak{grM}(R)$ (resp. $\in \mathfrak{grC}(R)$), $\mathrm{Hom}_R(M, N)$ is a graded module of all graded R-homomorphisms from M to N and that $\mathrm{Hom}_{\mathfrak{grM}(R)}(M, N)$ (resp. $\mathrm{Hom}_{\mathfrak{grC}(R)}(M, N)$) is the degree 0 part of $\mathrm{Hom}_R(M, N)$.

For the convenience we say that the category $\mathfrak{grC}(R)$ is **of finite representation type** if there are only a finite number of isomorphism classes of indecomposable graded CM modules up to degree shifting. We note that any arguments in previous chapters have their graded versions, so that the statements concerning $\mathfrak{C}(\hat{R})$ in preceeding chapters are, after a slight modification, all valid for the category $\mathfrak{grC}(R)$. For example the graded version of Theorem (4.22) can be stated as follows: If $\mathfrak{grC}(R)$ is of finite representation type, then R is a graded isolated singularity, by which we understand that each graded localization $R_{(\mathfrak{p})} = \{x/a|\ x \in R,\ a \in R - \mathfrak{p}$ is homogeneous$\}$ is regular if $\mathfrak{p}$ is a graded prime ideal with $\mathfrak{p} \neq \mathfrak{m}$.

We would like to prove the following

(15.1) PROPOSITION. $\mathfrak{grC}(R)$ *is of finite representation type if* $\mathfrak{C}(\widehat{R})$ *is.*

For any $M \in \mathfrak{grC}(R)$ we denote its $\mathfrak{m}$-adic completion by $\widehat{M}$. Likewise $\widehat{f}$ is the completion of a graded homomorphism f between graded modules. Thus we can define the functor from $\mathfrak{grC}(R)$ into $\mathfrak{C}(\widehat{R})$ by taking completion.

Proposition (15.1) now follows from the following

(15.2) LEMMA.
(15.2.1) *If M is an indecomposable graded R-module in* $\mathfrak{gr}\mathfrak{M}(R)$, *then $\widehat{M}$ is also indecomposable as an $\widehat{R}$-module.*
(15.2.2) *Let M and N be indecomposable objects in* $\mathfrak{gr}\mathfrak{M}(R)$. *If $\widehat{M} \simeq \widehat{N}$ in* $\mathfrak{M}(\widehat{R})$, *then M is isomorphic to N up to degree shifting.*

PROOF: (15.2.1): Write $E = \sum_{i \in \mathbb{Z}} E_i = \mathrm{End}_R(M)$ and $\mathfrak{N} = \sum_{i<0} E_i$. Note that $\mathfrak{N}$ is a nilpotent graded ideal of E. Since idempotents are split in the category $\mathfrak{gr}\mathfrak{M}(R)$, $E_0 = \mathrm{End}_{\mathfrak{gr}\mathfrak{M}(R)}(M)$ is a local ring. Note also that $\mathrm{End}_{\widehat{R}}(\widehat{M}) \simeq \widehat{E}$. Let $e(\neq 0)$ be an idempotent in $\widehat{E}$. We have to show that $e = 1$. Denote by $\bar{e}$ the class of e in $\widehat{E}/\mathfrak{N}\widehat{E}$. Since there is an isomorphism $\widehat{E}/\mathfrak{N}\widehat{E} \simeq (E/\mathfrak{N}E)\widehat{} = \prod_{i=0}^{\infty} E_i$, we can write $\bar{e} = e_r + e_{r+1} + e_{r+2} + \ldots$ $(r \geq 0, \ e_r \neq 0)$ where each e_i is in E_i. Comparing the terms of minimal degree in $\bar{e}^2 = \bar{e}$, we see that $e_r^2 = e_r$, hence we obtain $r = 0$ and $e_0 = 1$, as E_0 is a local ring. Since $\sum_{i>0} e_i \in \mathrm{rad}\,\widehat{E}$, this implies that $\bar{e}$ is a unit in $\widehat{E}/\mathfrak{N}\widehat{E}$. Therefore e is a unit in $\widehat{E}$ and thus $e = 1$ as required.

(15.2.2): Let $H_1 = \mathrm{Hom}_R(M, N)$ and $H_2 = \mathrm{Hom}_R(N, M)$. Since M and N are finitely generated R-modules, we see that

$$H_1 \otimes_R \widehat{R} \simeq \mathrm{Hom}_{\widehat{R}}(\widehat{M}, \widehat{N}), \qquad H_2 \otimes_R \widehat{R} \simeq \mathrm{Hom}_{\widehat{R}}(\widehat{N}, \widehat{M}).$$

By the assumption there are $f \in \mathrm{Hom}_{\widehat{R}}(\widehat{M}, \widehat{N})$ and $g \in \mathrm{Hom}_{\widehat{R}}(\widehat{N}, \widehat{M})$ with $f \cdot g = 1_{\widehat{N}}$ and $g \cdot f = 1_{\widehat{M}}$. Then, by the above isomorphisms, we have homogeneous elements $h_{ij} \in H_i$ and $a_{ij} \in \widehat{R}$ for $i = 1, 2$ such that $f = \sum_j h_{1j} \otimes a_{1j}$ and $g = \sum_j h_{2j} \otimes a_{2j}$. Thus $f \cdot g = 1_{\widehat{N}}$ implies $\sum_{jk} h_{1j} h_{2k} \otimes a_{1j} a_{2k} = 1_{\widehat{N}}$. Since $\mathrm{End}_R(N) \otimes_R \widehat{R} \simeq \mathrm{End}_{\widehat{R}}(\widehat{N})$ is a local ring by (15.2.1), we see that, for some j and k, $h_{1j} h_{2k} \otimes a_{1j} a_{2k}$ is a unit in this ring. Then $a_{1j} a_{2k}$ is a unit in $\widehat{R}$ and $h_{1j} h_{2k}$ is a graded automorphism on N. Thus N is isomorphic to a direct summand of M up to degree shifting. Since both M and N are indecomposable, this shows (15.2.2). ∎

This lemma shows that the set of classes of indecomposable graded CM modules over R under isomorphisms up to degree shifting is a subset of the set of isomorphism classes of indecomposable CM modules over $\widehat{R}$. Therefore the proposition follows.

We can prove that the converse of Proposition (15.1) is true as well. To do this we need several auxiliary results.

(15.3) DEFINITION. We say that a finitely generated module M over $\hat{R}$ is **gradable** if there is a finitely generated graded module X over R such that $M \simeq \widehat{X}$. Similarly an $\hat{R}$-homomorphism $f : N \to M$ of gradable modules is said to be a **gradable homomorphism** if there is a graded homomorphism of graded modules $g : Y \to X$ with a commutative diagram in $\mathfrak{C}(\hat{R})$:

$$\begin{array}{ccc} N & \xrightarrow{\simeq} & \hat{Y} \\ f \downarrow & & \downarrow \hat{g} \\ M & \xrightarrow{\simeq} & \widehat{X} \end{array}$$

(15.4) REMARK. Let K_R be the graded canonical module of the graded CM ring R. (For the definition of K_R, we understand (1.10) in the graded sense.) Then it is known and can be easily seen that the canonical module $K_{\hat{R}}$ of $\hat{R}$ is isomorphic to $\widehat{K_R}$, whence it is gradable. For a finitely generated graded R-module X, denote the graded canonical dual $\operatorname{Hom}_R(X, K_R)$ by X'. And for an $\hat{R}$-module M we also write $M' = \operatorname{Hom}_{\hat{R}}(M, K_{\hat{R}})$. Then by the above if $M = \widehat{X}$, then $M' = \widehat{X'}$. As a result, *the canonical dual of a gradable module over $\hat{R}$ is also gradable.* Similarly, for a gradeable $\hat{R}$-module M, the $\hat{R}$-dual $M^* = \operatorname{Hom}_{\hat{R}}(M, \hat{R})$ of M is also a gradable module.

Let X be a graded CM module over R and let $M = \widehat{X}$. Consider a syzygy of X as a graded R-module:

$$0 \longrightarrow Y \longrightarrow F_{n-1} \xrightarrow{f_{n-1}} F_{n-2} \longrightarrow \cdots \longrightarrow F_1 \xrightarrow{f_1} F_0 \longrightarrow X \longrightarrow 0,$$

where all F_i ($0 \le i < n$) are free R-modules and f_i ($1 \le i < n$) are graded R-homomorphisms. Then by the exactness of completion, it follows that $\hat{Y}$ is a syzygy of M as an $\hat{R}$-module. This implies that a syzygy module of a gradable module is also gradable. A similar argument to this shows that $\operatorname{tr}(M)$ is also gradable whenever M is gradable.

Recalling from (3.11) that AR translation is given by

$$\tau(M) = (\operatorname{syz}^d \operatorname{tr}(M))',$$

we conclude that $\tau(M)$ is gradable if M is. More precisely, if $M \simeq \widehat{X}$ for a gradable CM module X, then, denoting $\tau_{gr}(X) = (\operatorname{syz}^d \operatorname{tr}(M))'$ where syz and tr are taken in the graded sense, we have the equality:

(15.4.1) $$\tau(M) = \widehat{\tau_{gr}(X)}.$$

Furthermore we notice that *any direct summand of a gradable module is gradable.* In fact, if $M = \widehat{X}$ with X a graded module, then, decomposing X into a sum of graded indecomposable modules: $X = \sum_i X_i$, we see from (15.2.1) that $M = \sum_i \widehat{X_i}$ is a direct decomposition of M into indecomposable modules, hence that any summand of M is a sum of several $\widehat{X_i}$'s by the Krull- Schmidt theorem.

Recall that R is a **graded isolated singularity** if each graded localization $R_{(\mathfrak{p})}$ is regular for any graded prime $\mathfrak{p}(\neq \mathfrak{m})$. If this is the case, the graded version of (3.3) shows that any graded CM module over R is locally free, that is, for any graded prime ideal $\mathfrak{p}(\neq \mathfrak{m})$, $M_{(\mathfrak{p})}$ is $R_{(\mathfrak{p})}$-free.

(15.5) LEMMA. *Assume that R is a graded isolated singularity. Then, for any $X, Y \in \mathfrak{grC}(R)$ and for any positive integer n, there is a natural isomorphism:*

$$\operatorname{Ext}_R^n(X, Y) \simeq \operatorname{Ext}_{\widehat{R}}^n(\widehat{X}, \widehat{Y}).$$

PROOF: Since the completion is faithfully flat, there is a natural isomorphism of $\widehat{R}$-modules: $\operatorname{Ext}_R^n(X,Y)\widehat{} \simeq \operatorname{Ext}_{\widehat{R}}^n(\widehat{X}, \widehat{Y})$. Since the graded CM module X is locally free, $\operatorname{Ext}_R^n(X, Y)$ is an Artinian module for $n > 0$, in particular it is a complete module, i.e., $\operatorname{Ext}_R^n(X,Y)\widehat{} \simeq \operatorname{Ext}_R^n(X,Y)$. ∎

(15.6) LEMMA. *Assume that R is a graded isolated singularity and let M be an indecomposable gradable CM module over $\widehat{R}$.*
(15.6.1) *If M is not free, then there is an AR sequence ending in M:*

$$0 \longrightarrow \tau(M) \xrightarrow{q} E \xrightarrow{p} M \longrightarrow 0,$$

where $\tau(M)$, E are gradable modules and p, q are gradable homomorphisms.
(15.6.2) *If M is not isomorphic to the canonical module $K_{\widehat{R}}$, then there is an AR sequence starting from M:*

$$0 \longrightarrow M \xrightarrow{q} G \xrightarrow{p} \tau^{-1}(M) \longrightarrow 0,$$

where G, $\tau^{-1}(M)$ are gradable modules and p, q are gradable homomorphisms.

PROOF: By duality it is sufficient to prove (15.6.1). Note first that $\widehat{R}$ is an isolated singularity, since the defining ideal of singular locus of $\widehat{R}$ is gradable and it is $\widehat{\mathfrak{m}}$-primary by the assumption. Let X be a graded CM module over R such that $\widehat{X} \simeq M$. We showed (cf. (3.13)) that the AR sequence σ ending in M corresponds to the unique socle element in $\operatorname{Ext}_{\widehat{R}}^1(M, \tau(M))$. Since $\operatorname{Ext}_{\widehat{R}}^1(M, \tau(M)) \simeq \operatorname{Ext}_R^1(X, \tau_{gr}(X))$ by (15.4.1) and (15.5), its socle is in part of maximal degree in this graded module. Hence we can take a socle element as a homogeneous element in $\operatorname{Ext}_R^1(X, \tau_{gr}(X))$. Let $\rho : 0 \to \tau_{gr}(X) \to$

$Y \to X \to 0$ be a short exact sequence of graded CM modules corresponding to the homogeneous socle. Then the AR sequence ending in M is given by the completion of ρ. Hence all the modules and all the homomorphisms appearing in the AR sequence are gradable. ∎

(15.7) PROPOSITION. *Suppose that R is a graded isolated singularity. Let M and N be indecomposable CM modules over $\hat{R}$ and assume that there is an irreducible morphism $M \to N$ in $\mathfrak{C}(\hat{R})$.*
(15.7.1) *If N is nonfree and gradable, then M is gradable.*
(15.7.1)' *If M is not isomorphic to $K_{\hat{R}}$ and if M is gradable, then N is gradable.*
(15.7.2) *If M is nonfree and gradable, then N is gradable.*
(15.7.2)' *If N is not isomorphic to $K_{\hat{R}}$ and if N is gradable, then M is gradable.*

PROOF: (15.7.1): Since N is nonfree and gradable, it follows from (15.6.1) that there is an AR sequence
$$0 \longrightarrow \tau(N) \longrightarrow E \longrightarrow N \longrightarrow 0,$$
where all modules are gradable. Then by (2.12), M is a direct summand of E, hence it is also gradable.

(15.7.2): If $N \simeq R$, then there is nothing to prove. So we assume that N is nonfree. Then there is an AR sequence of the form:
$$0 \longrightarrow \tau(N) \longrightarrow E \oplus M \longrightarrow N \longrightarrow 0,$$
since there is an irreducble morphism from M to N. Thus, by (2.12), there is an irreducible morphism from $\tau(N)$ to M. Since M is nonfree and gradable, we can construct an AR sequence
$$0 \longrightarrow \tau(M) \longrightarrow L \longrightarrow M \longrightarrow 0,$$
consisting of gradable modules; see (15.6.1). Then $\tau(N)$ is a direct summand of L, hence it is gradable. Since $\tau(N) \not\simeq K_{\hat{R}}$, (15.6.2) shows that $N = \tau^{-1}(\tau(N))$ is also gradable.

(15.7.1)' and (15.7.2)' follow from (15.6) by a similar argument to the above. ∎

As a special case of this proposition we obtain:

(15.8) COROLLARY. *Suppose that R is a graded isolated singularity and that $\hat{R}$ is not Gorenstein. Let M and N be indecomposable CM modules over $\hat{R}$ and assume that there is an irreducible morphism $M \to N$ in $\mathfrak{C}(\hat{R})$. If one of M and N is gradable, then both are gradable.*

PROOF: Since $K_{\hat{R}} \not\simeq \hat{R}$ by the assumption, we can apply (15.7) to get the corollary. ∎

(15.9) NOTATION. We denote by Γ the AR quiver of the category $\mathfrak{C}(\widehat{R})$. Moreover Γ_{gr} denotes the subgraph of Γ consisting of all vertices of indecomposable gradable CM modules and all arrows connecting them.

With this notation, (15.8) can be stated as follows:

(15.10) COROLLARY. *Suppose that R is a graded isolated singularity and that $\widehat{R}$ is not Gorenstein. Then Γ_{gr} is a sum of several connected components of Γ.*

As we naturally expect, this corollary is true also in the case that $\widehat{R}$ is Gorenstein. To show this we need the following proposition.

(15.11) PROPOSITION. *Suppose that $\widehat{R}$ is a Gorenstein ring. Let M be a gradable $\widehat{R}$-module that is not necessarily CM and let n be a nonnegative integer. Then there is a gradable CM module L over $\widehat{R}$ and there is an epimorphism*

$$\mathrm{Hom}_{\widehat{R}}(\ ,L) \longrightarrow \mathrm{Ext}^n_{\widehat{R}}(\ ,M)$$

in the Auslander category $\mathrm{mod}(\mathfrak{C}(\widehat{R}))$.

PROOF: Take a graded R-module X with $\widehat{X} \simeq M$. As in (4.18) we prove the proposition by induction on $t = \dim(R) - \mathrm{depth}(X)$, where $\dim(R)$ (resp. $\mathrm{depth}(X)$) is the graded Krull dimension of R (resp. the graded depth of X). Note that $\dim(R) = \dim(\widehat{R})$ and $\mathrm{depth}(X) = \mathrm{depth}(M)$.

Suppose $t = 0$. Then X is a graded CM module. So, if $n = 0$, then it is enough to put $L = M$. Taking a graded free cover of the graded canonical dual X' of X, we have an exact sequence of graded modules: $0 \to Y \to F \to X' \to 0$, where F is free and Y is a graded CM module over R. Clearly its dual sequence is also exact:

$$0 \longrightarrow X \longrightarrow F' \longrightarrow Y' \longrightarrow 0.$$

We take the completion of this sequence to get the exact sequence of $\widehat{R}$-modules:

$$0 \longrightarrow M \longrightarrow \widehat{F'} \longrightarrow \widehat{Y'} \longrightarrow 0,$$

where we note that $\widehat{Y'}$ is also a gradable CM module over $\widehat{R}$. Then we obtain an exact sequence and isomorphisms in $\mathrm{mod}(\mathfrak{C}(\widehat{R}))$:

$$0 \longrightarrow \mathrm{Hom}_{\widehat{R}}(\ ,M) \longrightarrow \mathrm{Hom}_{\widehat{R}}(\ ,\widehat{F'}) \longrightarrow \mathrm{Hom}_{\widehat{R}}(\ ,\widehat{Y'}) \longrightarrow \mathrm{Ext}^1_{\widehat{R}}(\ ,M) \longrightarrow 0,$$

$$\mathrm{Ext}^n_{\widehat{R}}(\ ,M) \simeq \mathrm{Ext}^{n-1}_{\widehat{R}}(\ ,\widehat{Y'}) \quad (n > 1).$$

The first sequence shows the proposition in the case $n = 1$. For $n > 1$, the above isomorphism shows that the proof is reduced to the case $n - 1$. Hence the proposition is proved by induction on n.

Next suppose $t \geq 1$. Let $0 \to Z \to G \to X \to 0$ be an exact sequence of graded R-modules where G is free and Z is a graded module with $\operatorname{depth}(Z) = \operatorname{depth}(X) + 1$. Letting $N = \widehat{Z}$ and $P = \widehat{G}$, we have a short exact sequence $0 \to N \to P \to M \to 0$ and we may apply the induction hypothesis to N. Note that $\operatorname{Ext}^i_{\widehat{R}}(\ , P) = 0$ $(i > 0)$ as an object of $\operatorname{mod}(\mathfrak{C}(\widehat{R}))$, since $\widehat{R}$ is a Gorenstein ring and since P is a free $\widehat{R}$-module. Thus we have isomorphisms of functors:

$$\operatorname{Ext}^n_{\widehat{R}}(\ , M) \simeq \operatorname{Ext}^{n+1}_{\widehat{R}}(\ , N) \quad (n \geq 1).$$

This shows that, by the induction hypothesis, the proposition is true when $n \geq 1$. It remains to show in the case $n = 0$. From the above exact sequence, we have an exact sequence:

$$\operatorname{Hom}_{\widehat{R}}(\ , P) \longrightarrow \operatorname{Hom}_{\widehat{R}}(\ , M) \longrightarrow \operatorname{Ext}^1_{\widehat{R}}(\ , N) \longrightarrow 0.$$

By the induction hypothesis there is a gradable CM module Q such that an epimorphism $\operatorname{Hom}_{\widehat{R}}(\ , Q) \to \operatorname{Ext}^1_{\widehat{R}}(\ , N)$ exists in $\operatorname{mod}(\mathfrak{C}(\widehat{R}))$. Since $\operatorname{Hom}_{\widehat{R}}(\ , Q)$ is a projective object in the Auslander category $\operatorname{mod}(\mathfrak{C}(\widehat{R}))$ (see (4.8)), the above epimorphism can be lifted to a morphism $\operatorname{Hom}_{\widehat{R}}(\ , Q) \to \operatorname{Hom}_{\widehat{R}}(\ , M)$. Then putting $L = P \oplus Q$, we can construct an epimorphism $\operatorname{Hom}_{\widehat{R}}(\ , L) \to \operatorname{Hom}_{\widehat{R}}(\ , M)$. ∎

As a corollary of Proposition (15.11) we can show that a CM approximation of a gradable module over $\widehat{R}$ can be taken as a gradable CM module. More precisely we show:

(15.12) COROLLARY. *Suppose that $\widehat{R}$ is a Gorenstein ring. Let M be a gradable module in $\operatorname{gr}\mathfrak{M}(R)$. Then there exist a gradable CM module L over $\widehat{R}$ and a gradable homomorphism $f \in \operatorname{Hom}_{\widehat{R}}(L, M)$ that satisfy:*

(15.12.1) *Any homomorphism from any CM $\widehat{R}$-module N to M is a composition of f with some homomorphism from N to L.*

PROOF: Apply (15.11) to the case $n = 0$. ∎

Using this corollary we can show that (15.10) is true in general.

(15.13) PROPOSITION. *Let R be a graded isolated singularity. Then Γ_{gr} is a sum of several connected components of Γ.*

PROOF: By (15.10) we may assume that $\widehat{R}$ is a Gorenstein ring. Let $[M]$ be any vertex of Γ_{gr}. We have shown in (15.7) that if M is nonfree, then any vertex connected with

$[M]$ by an arrow in Γ belongs to Γ_{gr}. We have to show the same when $M \simeq R$. Since the maximal ideal $\widehat{\mathfrak{m}}$ of $\widehat{R}$ is a gradable module, it follows from (15.12) that there are a gradable CM module L and a gradable homomorphism $f : L \to \widehat{\mathfrak{m}}$ with the property (15.12.1).

Let N be an indecomposable CM module over $\widehat{R}$. If there is an irreducible morphism $g : N \to \widehat{R}$, then the image of g is in $\widehat{\mathfrak{m}}$, thus g is decomposed as $N \to L \xrightarrow{f} \widehat{\mathfrak{m}} \subset \widehat{R}$. Then by the definition of irreducible morphisms, N is isomorphic to a direct summand of L, and hence N is also gradable.

If there is an irreducible morphism $R \to N$, then the dual $N' \to R' = R$ is also an irreducible morphism and the above shows that N' is gradable, hence N is also gradable by (15.4). Thus the proposition is proved. ∎

Now we can prove the converse of (15.1).

(15.14) THEOREM. (Auslander-Reiten [15]) *Suppose that $R_0 = k$ is a perfect field. If $\mathfrak{gr}\mathfrak{C}(R)$ is of finite representation type, then so is $\mathfrak{C}(\widehat{R})$. And if this is the case, all CM modules on $\widehat{R}$ are gradable.*

PROOF: Since $\mathfrak{gr}\mathfrak{C}(R)$ is of finite representation type, the graded version of Theorem (4.22) shows that R is a graded isolated singularity. Hence by (15.13), Γ_{gr} is a sum of connected components of Γ. On the other hand, note that Γ_{gr} is a finite graph by the assumtion. Then Theorem (6.2) implies that $\Gamma = \Gamma_{gr}$. ∎

Chapter 16. CM modules on toric singularities

In this chapter we are interested in the representation type of toric singularities, and we shall give some examples of CM rings of dimension three, which are non-Gorenstein and of finite representation type.

In what follows k will denote an *algebraically closed field of characteristic* 0 and k^* denotes the multiplicative group of all non-zero elements in k.

Recall that an **algebraic torus** T of dimension m is a direct product $(k^*)^m$. Suppose that the torus T acts rationally on a vector space V of dimension n. Then it can be seen by complete reducibility of torus-action that there exist a basis $\{x_1, x_2, \ldots, x_n\}$ of V and integers a_{ij} ($1 \leq i \leq n$, $1 \leq j \leq m$) such that the action is given by $t \cdot x_i = t_1^{a_{i1}} t_2^{a_{i2}} \cdots t_m^{a_{im}} x_i$ ($1 \leq i \leq n$) for $t = (t_1, t_2, \ldots, t_m) \in T$. In particular, the action is completely determined by the integer matrix $A = (a_{ij})$. This action can be extended to the symmetric algebra $S = S(V) = k[x_1, x_2, \ldots, x_n]$ of V and we denote the invariant subring of S by $R(A)$, i.e.

$$R(A) = \{f \in S | \; f(t \cdot x_1, t \cdot x_2, \ldots, t \cdot x_n) = f(x_1, x_2, \ldots, x_n) \text{ for any } t \in T\}.$$

This ring may be written in the following manner:
If H is the sub-semigroup of $\mathbb{N}^{(n)}$ consisting of $\alpha = (\alpha_1, \alpha_2, \ldots, \alpha_n)$ with $\sum_i \alpha_i a_{ij} = 0$ ($1 \leq j \leq m$), then $R(A)$ is the semigroup ring $k[x^\alpha| \; \alpha \in H]$, where x^α denotes $x_1^{\alpha_1} x_2^{\alpha_2} \cdots x_n^{\alpha_n}$.

We can make the ring $R(A)$ a $\mathbb{Z}$-graded ring by defining degree as follows:

(16.1.1) $$\deg(x^\alpha) = \sum_{i=1}^n \alpha_i.$$

Writing $R_+(A)$ for the maximal ideal generated by all homogeneous elements of positive degree, we define $\widehat{R}(A)$ to be the $R_+(A)$-adic completion of $R(A)$ and call it a **toric singularity**. Note that $\widehat{R}(A) = k[[x^\alpha| \; \alpha \in H]]$ with the above notation.

Of most importance is the following fact proved by Hochster [40]:

(16.1) PROPOSITION. $R(A)$ and $\hat{R}(A)$ are both CM rings.

By this fact it is natural to ask when $\hat{R}(A)$ is of finite representation type. By virtue of (15.1) and (15.14) the problem can be reduced to asking about the categroy $\mathfrak{grC}(R(A))$ of graded CM modules over $R(A)$. For convenience we make the following

(16.2) DEFINITION. An integer matrix A is said to be **of finite type** if the toric singularity $\hat{R}(A)$ is of finite representation type, otherwise it is **of infinite type**.

For example $A = {}^t(1, 1, -1, -1)$ is of finite type, since

$$\hat{R}(A) = k[[x_1x_3, x_1x_4, x_2x_3, x_2x_4]] \simeq k[[x, y, z, w]]/(xw - yz)$$

is a ring of simple singularity of type (A_1), while $B = {}^t(2, 1, -2, -1)$ is not, because

$$\hat{R}(B) = k[[x_1x_3, x_1x_4^2, x_2^2 x_3, x_2x_4]] \simeq k[[x, y, z, w]]/(xw^2 - yz).$$

(16.3) REMARK.
(16.3.1) Let A be an integer matrix and let B be a matrix obtained from A by successively permuting rows, making elementary transformation on columns and multiplying by a nonzero integer. Then $R(A)$ is isomorphic to $R(B)$ as a k-algebra, hence A is of finite type if and only if B is.
(16.3.2) Let A be an integer matrix, one of whose rows is null, and let B be a matrix obtained from A by deleting a null row. Then A is of finite type if and only if $R(B)$ is a polynomial ring over k.

Actually, note that under the assumption, $R(A) \simeq R(B)[x]$ (a polynomial ring over $R(B)$). Therefore the claim is immediate from the fact that a ring of finite representation type has only an isolated singularity, (4.22).
(16.3.3) Let A be an integer matrix of size $n \times 1$. If $n \leq 3$, then A is of finite type.

This is just a restatement of the fact that a quotient singularity of dimension 2 is of finite representation type, (10.14). In fact, by (16.3.1), we may assume that $A = {}^t(-a, b, c)$ with a, b and c positive integers. Then consider a cyclic group $\langle \sigma \rangle$ of order a and define the action of σ on the formal power series ring $S_1 = k[[y_1, y_2]]$ by $\sigma(y_1) = \zeta^b y_1$, $\sigma(y_2) = \zeta^c y_2$, where ζ is a primitive a-th root of unity. It is easy to see that $\hat{R}(A)$ is isomorphic to the ring of invariants $S_1^{\langle \sigma \rangle}$.

Adding to the above we show:

(16.4) PROPOSITION. *Let A be an integer matrix of any size and let B be a matrix obtained from A by deleting one of its rows. Then B is of finite type if A is.*

PROOF: Assume that $A = (a_{ij})$ is an $(n \times m)$-matrix. We may assume by (16.3.1) that one gets B by deleting the first row of A. First we claim the following:

(16.4.1) There is a natural embedding of k-algebras: $\hat{R}(B) \to \hat{R}(A)$, which has an algebra-retraction.

To prove this, let $H(A) = \{\alpha \in \mathbb{N}^{(n)} |\ \alpha A = 0\}$ and $H(B) = \{\beta \in \mathbb{N}^{(n-1)} |\ \beta B = 0\}$. We may consider $H(B)$ as a sub-semigroup of $H(A)$ by regarding each β as $(0, \beta)$ in $\mathbb{N}^{(n)}$. By this we have an embedding of $\hat{R}(B)$ into $\hat{R}(A)$. For example, if $A = {}^t(1, 2, -1, -1)$ and $B = {}^t(2, -1, -1)$, then $\hat{R}(B) = k[[x_2 x_3^2, x_2 x_3 x_4, x_2 x_4^2]] \subset \hat{R}(A) = k[[x_1 x_3, x_1 x_4, x_2 x_3^2, x_2 x_3 x_4, x_2 x_4^2]]$. In this example, if $\mathfrak{a}$ is an ideal of the ring $\hat{R}(A)$ generated by all monomials containing the variable x_1, then $\hat{R}(B) \simeq \hat{R}(A)/\mathfrak{a}$ thus the ring extension has a retraction. This works in general setting. Actually, we define a k-algebra mapping $\pi : R(A) \to R(B)$ by sending a monomial $x^\alpha = x_1^{\alpha_1} x_2^{\alpha_2} \cdots x_n^{\alpha_n}$ to $x_2^{\alpha_2} x_3^{\alpha_3} \cdots x_n^{\alpha_n}$ if $\alpha_1 = 0$, and otherwise 0. It can easily be seen that π is well-defined and it gives a retraction of the algebra extension $R(B) \subset R(A)$. Taking the completion of this, we show (16.4.1).

Secondly we notice that the dimension of $\hat{R}(A)$ is equal to the dimension of the $\mathbb{Q}$-vector space $U(A)$ spanned by all integral vectors in $H(A)$; see [40].

Note that $U(B) = U(A) \cap \{\alpha_1 = 0\}$, hence either $U(A) = U(B)$ or $\dim(\hat{R}(A)) = \dim(\hat{R}(B)) + 1$. In the first case, there is nothing to prove, because $\hat{R}(A) = \hat{R}(B)$. We thus may assume the equality $\dim(\hat{R}(A)) = \dim(\hat{R}(B)) + 1$.

By the claim (16.4.1), the proposition is a direct consequence of the following more general result:

(16.5) THEOREM. *Let $R \subset R'$ be a ring extension of normal analytic local CM domains. Suppose that the extension has a ring retraction and that $\dim(R') = \dim(R) + 1$. Then R is of finite representation type if R' is.*

PROOF: If $\dim(R) \leq 1$, the theorem is obviously true, since then, R is a regular local ring. So we assume that $\dim(R) \geq 2$. Note that R' is an isolated singularity, because it is of finite representation type. Since the extension $R \subset R'$ has a ring retraction, we may write $R = R'/\mathfrak{P}$, where $\mathfrak{P}$ is a prime ideal of R' of height one. Letting $\mathfrak{P}^{-1} = \{x \in Q(R) |\ x\mathfrak{P} \subset R'\}$, we consider the exact sequence of R'-modules:

(*) $$0 \longrightarrow R' \longrightarrow \mathfrak{P}^{-1} \overset{\pi}{\longrightarrow} U \longrightarrow 0,$$

where $U = \mathfrak{P}^{-1}/R'$. It follows from the definition that $\mathfrak{P} U = 0$, hence U is a module over $R = R'/\mathfrak{P}$. We notice the following:

(16.5.1) For any $\mathfrak{Q} \in \mathrm{Spec}(R')$ that is distinct from the maximal ideal of R', $\mathfrak{P}^{-1} R'_{\mathfrak{Q}}$ is a free module over $R'_{\mathfrak{Q}}$.

(16.5.2) For any $\mathfrak{q} \in \operatorname{Spec}(R)$ that is distinct from the maximal ideal of R, $\pi \otimes_{R'} R_{\mathfrak{q}}$ is an isomorphism of $R_{\mathfrak{q}}$-modules.

In fact, $R'_{\mathfrak{Q}}$ is a regular local ring, hence a UFD in (16.5.1). Thus $\mathfrak{P} R'_{\mathfrak{Q}}$ is a principal ideal, and so $\mathfrak{P}^{-1} R'_{\mathfrak{Q}} = (\mathfrak{P} R'_{\mathfrak{Q}})^{-1}$ is also principal, from which (16.5.1) follows. To show (16.5.2), let $\mathfrak{Q}$ be the prime ideal of R' with $\mathfrak{Q}/\mathfrak{P} = \mathfrak{q}$. Then $\mathfrak{P}\mathfrak{P}^{-1} R'_{\mathfrak{Q}} = R'_{\mathfrak{Q}}$ by the above argument, hence $\pi \otimes_{R'} R_{\mathfrak{q}} : \mathfrak{P}^{-1} R'_{\mathfrak{Q}} / \mathfrak{P}\mathfrak{P}^{-1} R'_{\mathfrak{Q}} \to \mathfrak{P}^{-1} R'_{\mathfrak{Q}} / R'_{\mathfrak{Q}}$ is an isomorphism.

From (16.5.1) and (16.5.2) we can show:

(16.5.3) For any finitely generatated R-module M, there is an isomorphism of R-modules:

$$M^* \simeq \operatorname{Tor}_1^{R'}(U, M)^*,$$

where ( )* indicates the dual by R.

Indeed, regarding M as an R'-module via $R' \to R$, we have an exact sequence of R-modules from (*):

$$0 \to \operatorname{Tor}_1^{R'}(\mathfrak{P}^{-1}, M) \to \operatorname{Tor}_1^{R'}(U, M) \xrightarrow{\varphi} M \to \mathfrak{P}^{-1} \otimes_{R'} M \xrightarrow{\pi \otimes M} U \otimes_{R'} M \to 0.$$

From (16.5.1) we see that $\operatorname{Tor}_1^{R'}(\mathfrak{P}^{-1}, M)$ is an R-module of finite length. Moreover the kernel of the map $\pi \otimes_{R'} M = (\pi \otimes_{R'} R) \otimes_R M$ is also of finite length by (16.5.2). Thus φ in the sequence is an isomorphism when localized at any prime ideal of R that is different from the maximal ideal. Since R is a normal domain of dimension ≥ 2, this implies that φ^* is an isomorphism, hence (16.5.3).

We finally claim the following:

(16.5.4) For any indecomposable CM modules M over R, there is an indecomposable CM module P over R' such that $\operatorname{rank}_R(P \otimes_{R'} U) \geq \operatorname{rank}_R(M)$ holds.

If this is true, then we are through. For, if R were not of finite representation type, then, by (6.4), the set $\{\operatorname{rank}_R(M) |\ M$ is an indecomposable CM module over $R\}$, hence $\{\operatorname{rank}_R(P \otimes_{R'} U) |\ P$ is an indecomposable CM module over $R'\}$, would have no bound and R' would be of infinite representation type.

To prove (16.5.4) let M be an indecomposable CM module over R. We consider the first syzygy L of M as an R'-module:

$$0 \longrightarrow L \longrightarrow R'^{(n)} \xrightarrow{p'} M \longrightarrow 0,$$

where $\operatorname{depth}_{R'}(L) = \operatorname{depth}_{R'}(M) + 1 = \dim(R')$, so L is a CM module over R'. Letting

$p = p' \otimes_{R'} R$ and $N = \text{Ker}(p)$, we now have the commutative diagram with exact rows:

$$\begin{array}{ccccccccc} 0 & \longrightarrow & L & \longrightarrow & R'^{(n)} & \xrightarrow{p'} & M & \longrightarrow & 0 \\ & & {\scriptstyle g}\downarrow & & {\scriptstyle f}\downarrow & & \| & & \\ 0 & \longrightarrow & N & \longrightarrow & R^{(n)} & \xrightarrow{p} & M & \longrightarrow & 0, \end{array}$$

where f is a natural projection and $g = f|_L$. If i is a natural inclusion of $R^{(n)}$ into $R'^{(n)}$, and if j is its restriction on N, then the image of j lies in L and we have $g \cdot j = 1_N$. This shows that $g \otimes_{R'} U : L \otimes_{R'} U \to N \otimes_{R'} U = N \otimes_R U$ is a split epimorphism of R-modules. Note from the above diagram that the kernel of $g \otimes_{R'} U$ is isomorphic to $\text{Tor}_1^{R'}(M, U)$. Consequently $\text{Tor}_1^{R'}(M, U)$ is a direct summand of $L \otimes_{R'} U$ as an R-module. Therefore $M^* \simeq \text{Tor}_1^{R'}(M, U)^*$ is also a direct summand of $(L \otimes_{R'} U)^*$. Notice that, since M is reflexive, M^* is an indecomposable module as well as M. Thus, by the Krull-Schmidt theorem, there is an indecomposable R'-summand P of L such that M^* is an R-summand of $(P \otimes_{R'} U)^*$, in particular,

$$\text{rank}_R(M) = \text{rank}_R(M^*) \leq \text{rank}_R((P \otimes_{R'} U)^*) = \text{rank}_R(P \otimes_{R'} U).$$

This proves (16.5.4), hence the theorem. ∎

In the rest of this chapter we restrict ourselves to situation involving only one-dimensional tori, i.e. $m = 1$ with the notation in the beginning of this chapter. In this case, the corresponding integer matrices are of size $n \times 1$ for some n.

(16.6) NOTATION. Suppose we are given an integer matrix $A = {}^t(a_1, a_2, \ldots, a_n)$. Without loss of generality, we may assume that the greatest common divisor d of $\{a_1, a_2, \ldots, a_n\}$ is 1. (If $d > 1$, then, considering the matrix $B = {}^t(a_1/d, a_2/d, \ldots, a_n/d)$, we clearly have $R(A) = R(B)$.) Then we can make the polynomial ring $S = k[x_1, x_2, \ldots, x_n]$ into a $\mathbb{Z}$-graded ring by the rule:

$$\deg(x_i) = a_i \qquad (1 \leq i \leq n).$$

For $c \in \mathbb{Z}$, denote by S_c the degree c part in S. Clearly, $S = \sum_{c \in \mathbb{Z}} S_c$ and $S_0 = R(A)$. For a graded S-module $M = \sum_{n \in \mathbb{Z}} M_n$ and for an integer c, we denote the shifted module by $M(c)$, that is, $M(c)_{c'} = M_{c+c'}$ for any $c' \in \mathbb{Z}$.

Let L be the quotient field of $R(A)$ and let Q be the total graded ring of quotients of S. Since $\{a_1, a_2, \ldots, a_n\}$ generates (1) as an ideal of $\mathbb{Z}$, we can find an element $t \in Q$ of degree 1. Then it is easy to see that $Q = L[t, t^{-1}]$. For any $c \in \mathbb{Z}$, since $S_c t^{-c} \subset L$, S_c is a (finitely generated) $R(A)$-module of rank one, hence S_c is isomorphic to a fractional ideal of $R(A)$.

For any finitely generated $\mathbb{Z}$-graded S-modules M, N and for any integers a, b, we note that there is a natural mapping of R-modules:

(16.7.1) $$\rho : \mathrm{Hom}_S(M, N)_{a-b} \longrightarrow \mathrm{Hom}_{R(A)}(M_b, N_a),$$

defined by restricting any homomorphism in $\mathrm{Hom}_S(M, N)_{a-b}$ on M_b. Unfortunately ρ may not be bijective in general. However, we can show the following:

(16.7) LEMMA. *Let a, b be integers. Suppose that S_b generates an ideal of S of height ≥ 2, i.e. $\mathrm{ht}(S_b S) \geq 2$. Then the map $\rho : \mathrm{Hom}_S(S, S)_{a-b}(\simeq S_{a-b}) \to \mathrm{Hom}_{R(A)}(S_b, S_a)$ is an isomorphism.*

PROOF: Let g be an element of $\mathrm{Hom}_S(S, S)_{a-b}$ and suppose that $g(S_b) = 0$. Then taking a nonzero element z in S_b, we show that $zg(1) = g(z) = 0$ and that $g(1) = 0$, since S is an integral domain. Thus $g(w) = wg(1) = 0$ for any $w \in S$, showing the injectivity of ρ.

To show that ρ is surjective, let $f : S_b \to S_a$ be an $R(A)$-homomorphism. Since $S_a \subset Lt^a$ and $S_b \subset Lt^b$, $f \otimes_{R(A)} L : Lt^b \to Lt^a$ is a multiplication map by zt^{a-b} for some $z \in L$, hence f is also a multiplication by zt^{a-b}. We have to show that $zt^{a-b} \in S_{a-b}$. For this, consider the S-homomorphism:

$$(f \otimes_{R(A)} S)^{**} : (S_b \otimes_{R(A)} S)^{**} \longrightarrow (S_a \otimes_{R(A)} S)^{**},$$

where ( )* denotes the dual by S. Note here that $(S_a \otimes_{R(A)} S)^{**} \simeq (S_a S)^{**} \subseteq S$ and $(S_b \otimes_{R(A)} S)^{**} \simeq (S_b S)^{**} = S$, because $\mathrm{ht}(S_b S) \geq 2$. By these isomorphisms, $(f \otimes_{R(A)} S)^{**}$ is also a multiplication by zt^{a-b} and is an element in $\mathrm{Hom}_S((S_b S)^{**}, (S_a S)^{**}) \simeq (S_a S)^{**} \subseteq S$. Hence $zt^{a-b} \in S \cap Lt^{a-b} = S_{a-b}$ as desired. ∎

We quote a fact from Stanley's theory which enables us to know the CM property for ideals S_c.

(16.8) PROPOSITION. (Stanley [62,Chap.1, 7.8]) *Let*

$$A = {}^t(a_1, a_2, \ldots, a_p, b_1, b_2, \ldots, b_q)$$

where $a_i > 0$ $(1 \leq i \leq p)$, $b_j < 0$ $(1 \leq j \leq q)$ and $p + q \geq 3$. If c is a positive (resp. negative) integer, then S_c is a graded CM module over $R(A)$ if and only if there are no

$(\alpha_1, \alpha_2, \ldots, \alpha_p, \beta_1, \beta_2, \ldots, \beta_q) \in \mathbb{Z}^{(p+q)}$ with $\alpha_i \geq 0$ (resp. $\alpha_i < 0$), $1 \leq i \leq p$, $\beta_j < 0$ (resp. $\beta_j \geq 0$), $1 \leq j \leq q$, and $\sum_{i=1}^{p} \alpha_i a_i + \sum_{j=1}^{q} \beta_j b_j = c$.

For example, if $A = {}^t(2, -1, -1, -1)$, then S_c is a CM $R(A)$-module if and only if $c = -1, 0, 1, 2$. We shall show that this matrix is of finite type; see (16.10). Before this we note the following, which is useful for determining AR quivers.

(16.9) LEMMA. *Let T be an analytic CM local ring with perfect residue field and let Γ° be a finite subgraph of the AR quiver of $\mathfrak{C}(T)$ which is closed under AR sequences and which contains the classes of the free module and of the canonical module. Suppose that T is an isolated singularity and non-Gorenstein. Then Γ° is the whole quiver, and hence T is of finite representation type.*

Here we say that Γ° is **closed under AR sequences** provided that all the classes of indecomposable modules appearing in an AR sequence $0 \to N \to E \to M \to 0$ belong to Γ° if one of the classes of M and N does.

PROOF: Denote by K the canonical module of T. By virtue of (6.2), it is enough to show that Γ° is a connected component of the AR quiver of T. Since Γ° is already closed under AR sequences, it suffices to show that there are no irreducible morphisms $X \to T$ or $K \to X$ with $[X] \notin \Gamma^\circ$. We show below that if there is an irreducible morphism $X \to T$ (resp. $K \to X$), then $[X] \in \Gamma^\circ$.

If $X \simeq K$ (resp. $\simeq T$), there is nothing to prove, since the class of K (resp. T) is in Γ°. Assume that X is not isomorphic to K (resp. T). Then, since T is an isolated singularity, there is an AR sequence: $0 \to X \to E \to \tau^{-1}(X) \to 0$ (resp. $0 \to \tau(X) \to E \to X \to 0$); cf. (3.2). Therefore we see from (2.12) and (2.12)' that T (resp. K) is a direct summand of E and that there is an irreducible morphism from T to $\tau^{-1}(X)$ (resp. from $\tau(X)$ to K). Since T is not isomorphic to K, we consequently have the AR sequence of the form: $0 \to T \to \tau^{-1}(X) \oplus G \to \tau^{-1}(T) \to 0$ (resp. $0 \to \tau(K) \to \tau(X) \oplus G \to K \to 0$). Therefore $\tau^{-1}(X)$ (resp. $\tau(X)$), hence X, lies in Γ°, since Γ° is closed under AR sequences. ∎

(16.10) PROPOSITION. (Auslander-Reiten [14])

The integer matrix $A = {}^t(2, -1, -1, -1)$ is of finite type.

Note that

$$\widehat{R}({}^t(2, -1, -1, -1)) = k[[x_1 x_2^2, x_1 x_2 x_3, x_1 x_3^2, x_1 x_3 x_4, x_1 x_4^2, x_1 x_2 x_4]],$$

which is isomorphic to the subring $k[[X^2, XY, Y^2, YZ, Z^2, ZX]]$ of $k[[X, Y, Z]]$.

PROOF: For simplicity, write R (resp. $\widehat{R}$) instead of $R(A)$ (resp. $\widehat{R}(A)$). Note that $\widehat{R}$ is an isolated singularity of dimension 3 and is non-Gorenstein.

As remarked after (16.8), S_{-1}, S_0, S_1 and S_2 are graded CM modules over R. Notice that

$$S_{-1} = (x_2, x_3, x_4)R, \quad S_0 = R, \quad S_1 = (x_1x_2, x_1x_3, x_1x_4)R \quad \text{and} \quad S_2 = (x_1)R.$$

Hence $\widehat{S_{-1}} \simeq \widehat{S_1}$ and $\widehat{S_0} \simeq \widehat{S_2}$ as $\hat{R}$-modules. Furthermore it is easy to see that $\widehat{S_{-1}}$ is isomorphic to the canonical module of $\hat{R}$.

First of all, we compute the AR translation of $\widehat{S_{-1}}$. Recall that $\tau = (\text{syz}_{\hat{R}}^3 \operatorname{tr}(\ ))'$, (3.11), and that the right hand side is isomorphic to $(\text{syz}_{\hat{R}}^1 (\ )^*)'$ where $(\ )^*$ means the $\hat{R}$-dual, since $(\ )^* \simeq \text{syz}_{\hat{R}}^2 \operatorname{tr}(\ )$. Note that $(\widehat{S_{-1}})^* = \operatorname{Hom}_R(S_{-1}, R)\widehat{\ } \simeq \widehat{S_1} \simeq \widehat{S_{-1}}$ by (16.7). To compute the first syzygy of $\widehat{S_{-1}}$ we take the Koszul complex over S:

$$(16.10.1) \quad 0 \longrightarrow S(3) \xrightarrow{\begin{pmatrix} x_4 \\ -x_3 \\ x_2 \end{pmatrix}} S(2)^{(3)} \xrightarrow{\begin{pmatrix} x_3 & x_4 & 0 \\ -x_2 & 0 & x_4 \\ 0 & -x_2 & -x_3 \end{pmatrix}} S(1)^{(3)} \xrightarrow{(x_2\ x_3\ x_4)} S.$$

Taking the degree -1 part in this sequence, we obtain an exact sequence of graded R-modules:

$$(16.10.2) \quad 0 \longrightarrow S_2 \xrightarrow{\begin{pmatrix} x_4 \\ -x_3 \\ x_2 \end{pmatrix}} S_1^{(3)} \xrightarrow{\begin{pmatrix} x_3 & x_4 & 0 \\ -x_2 & 0 & x_4 \\ 0 & -x_2 & -x_3 \end{pmatrix}} R^{(3)} \xrightarrow{(x_2\ x_3\ x_4)} S_{-1} \longrightarrow 0.$$

We define a CM $\hat{R}$-module M by the exact sequence:

$$(16.10.3) \quad 0 \longrightarrow M \longrightarrow \hat{R}^{(3)} \xrightarrow{(x_2\ x_3\ x_4)} \widehat{S_{-1}} \longrightarrow 0.$$

It is easy to see that M is an indecomposable module of rank 2. From (16.10.2) we have an exact sequence:

$$(16.10.4) \quad 0 \longrightarrow \widehat{S_2} \xrightarrow{\begin{pmatrix} x_4 \\ x_3 \\ x_2 \end{pmatrix}} \widehat{S_1}^{(3)} \longrightarrow M \longrightarrow 0.$$

By taking the canonical dual of this, we see that the following is also exact:

$$0 \longrightarrow M' \longrightarrow \operatorname{Hom}_R(S_1^{(3)}, S_{-1})\widehat{\ } \xrightarrow{(x_4\ x_3\ x_2)} \operatorname{Hom}_R(S_2, S_{-1})\widehat{\ }.$$

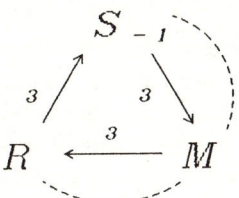

Figure (16.10.5).

Comparing this with (16.10.3), we have $M' \simeq M$, since $\operatorname{Hom}_R(S_1, S_{-1})\widehat{} \simeq \operatorname{Hom}_R(S_{-1}, S_{-1})\widehat{} \simeq \widehat{R}$ and $\operatorname{Hom}_R(S_2, S_{-1})\widehat{} \simeq \operatorname{Hom}_R(R, S_{-1})\widehat{} \simeq \widehat{S_{-1}}$ by (16.7). As a result, $\tau(\widehat{S_{-1}}) \simeq (\operatorname{syz}^1_{\widehat{R}}(\widehat{S_{-1}}))' \simeq M' \simeq M$.

On the other hand, from (16.10.4), the following sequence is exact:

$$0 \longrightarrow M^* \longrightarrow \widehat{S_{-1}}^{(3)} \xrightarrow{(x_4 \ x_3 \ x_2)} \widehat{S_{-2}},$$

since $(\widehat{S_2})^* \simeq \widehat{R}^* \simeq \widehat{R} \simeq \widehat{S_2}$ and $(\widehat{S_1})^* \simeq (\widehat{S_{-1}})^* \simeq \widehat{S_1} \simeq \widehat{S_{-1}}$ by (16.7). Hence, taking the degree -2 part in (16.10.1) and completing it, we see that the sequence

$$0 \longrightarrow \widehat{S_1} \xrightarrow{\begin{pmatrix} x_4 \\ -x_3 \\ x_2 \end{pmatrix}} \widehat{R}^{(3)} \longrightarrow M^* \longrightarrow 0.$$

is also exact. Therefore $\tau(M) \simeq (\operatorname{syz}^1_{\widehat{R}}(M^*))' \simeq (\widehat{S_1})' \simeq (\widehat{S_{-1}})' \simeq \widehat{R}$.

Applying the functor $\operatorname{Hom}_{\widehat{R}}(\widehat{S_{-1}}, \ )$ to (16.10.3), we have an exact sequence

$$\widehat{S_1}^{(3)} \xrightarrow{(x_2 \ x_3 \ x_4)} \widehat{R} \longrightarrow \operatorname{Ext}^1_{\widehat{R}}(\widehat{S_{-1}}, M),$$

and one can show that $1 \in \widehat{R}$ is sent to the socle element of $\operatorname{Ext}^1_{\widehat{R}}(\widehat{S_{-1}}, M)$ by the second mapping in the sequence. We, thus, conclude from (3.13) that the sequence (16.10.3) is the AR sequence ending in $\widehat{S_{-1}}$. Taking the canonical dual, we also show that the following is the AR sequence ending in M:

$$0 \longrightarrow \widehat{R} \xrightarrow{\begin{pmatrix} x_2 \\ x_3 \\ x_4 \end{pmatrix}} \widehat{S_{-1}}^{(3)} \longrightarrow M \longrightarrow 0.$$

Thus we have shown that Figure (16.10.5) is part of the AR quiver of $\mathfrak{C}(\widehat{R})$ which is closed under AR sequences. Hence Lemma (16.9) implies that it is the whole quiver. ∎

(16.11) REMARK. Auslander and Reiten showed that the ring in (16.10) is the unique CM ring of finite representation type that is an invariant subring by a finite group and has dimension ≥ 3. See [14] for further discussion.

In terms of toric singularities, their result shows that an integer matrix $A = {}^t(a, -b_1, -b_2, \ldots, -b_r)$ $(a > 0, b_i > 0, 1 \leq i \leq r)$ is of finite type if and only if it satisfies one of the following conditions:

(i) there exists an integer j $(1 \leq j \leq r)$ such that $b_i \equiv 0 \pmod{a}$ for any i but j;

(ii) $r \leq 2$;

(iii) $r = 3$, a is even and $b_i \equiv a/2 \pmod{a}$ for any i.

In fact, the same argument as in (16.3.3) shows that $\widehat{R}(A)$ is a cyclic quotient singularity and each condition above corresponds respectively to the case (i) $\widehat{R}(A)$ being regular, (ii) $\widehat{R}(A)$ having dimension ≤ 2 and (iii) $\widehat{R}(A)$ being the ring given in (16.10).

We give another example of a non-Gorenstein CM ring of dimension 3 that is of finite representation type.

(16.12) PROPOSITION. (Auslander-Reiten [14])
The integer matrix $A = {}^t(2, 1, -1, -1)$ is of finite type.

Here in this case,

$$\widehat{R}({}^t(2,1,-1,-1)) = k[[x_1 x_3^2, x_1 x_3 x_4, x_1 x_4^2, x_2 x_3, x_2 x_4]]$$
$$\simeq k[[X, Y, Z, U, V]]/(XZ - Y^2, XV - YU, YV - ZU).$$

PROOF: Denote $R = R(A)$ and $\widehat{R} = \widehat{R}(A)$. Notice that $\widehat{R}$ is an isolated singularity of dimension 3 and is non-Gorenstein.

By (16.8) the ideals S_{-2}, S_{-1}, S_0 and S_1 are graded CM modules over R. Note that

$$S_{-2} = (x_3^2, x_3 x_4, x_4^2)R, \quad S_{-1} = (x_3, x_4)R, \quad S_0 = R \quad \text{and} \quad S_1 = (x_1 x_3, x_1 x_4, x_2)R.$$

One can easily see that these are all non-isomorphic and that $\widehat{S_{-1}}$ is the canonical module of $\widehat{R}$. Taking the degree -1 part in the Koszul complex

$$0 \longrightarrow S(2) \xrightarrow{\begin{pmatrix} x_4 \\ -x_3 \end{pmatrix}} S(1)^{(2)} \xrightarrow{(x_3 \ x_4)} S \longrightarrow S/(x_3, x_4)S \longrightarrow 0,$$

we have an exact sequence of $\hat{R}$-modules:

$$0 \longrightarrow \widehat{S_1} \xrightarrow{\binom{x_4}{-x_3}} \hat{R}^{(2)} \xrightarrow{(x_3\ x_4)} \widehat{S_{-1}} \longrightarrow 0.$$

Hence, by (16.7), $\tau(\widehat{S_1}) \simeq (\mathrm{syz}^1_{\hat{R}}(\widehat{S_1}^*))' \simeq (\mathrm{syz}^1_{\hat{R}}(\widehat{S_{-1}}))' \simeq (\widehat{S_1})' \simeq \widehat{S_{-2}}$.
Likewise we obtain an exact sequence $0 \to \widehat{S_{-2}} \to \hat{R}^{(2)} \to \widehat{S_2} \to 0$ from

$$0 \longrightarrow S(-4) \xrightarrow{\binom{x_2^2}{-x_1}} S(-2)^{(2)} \xrightarrow{(x_1\ x_2^2)} S \longrightarrow S/(x_1, x_2^2)S \longrightarrow 0,$$

since $S_2 = (x_1, x_2^2)R$. Therefore $\tau(\widehat{S_{-2}}) \simeq (\mathrm{syz}^1_{\hat{R}}(\widehat{S_{-2}}^*))' \simeq (\mathrm{syz}^1_{\hat{R}}(\widehat{S_2}))' \simeq (\widehat{S_{-2}})' \simeq \widehat{S_1}$.
To compute $\tau(\widehat{S_{-1}})$ we take an exact sequence of graded S-modules:

$$0 \longrightarrow S(-1) \xrightarrow{\begin{pmatrix} x_2 \\ -x_4 \\ x_3 \end{pmatrix}} S \oplus S(-2)^{(2)} \xrightarrow{\begin{pmatrix} x_4 & x_2 & 0 \\ -x_3 & 0 & x_2 \\ 0 & -x_1x_3 & -x_1x_4 \end{pmatrix}} S(-1)^{(3)} \xrightarrow{(x_1x_3\ x_1x_4\ x_2)} S$$

Let M be the reduced first syzygy of the $\hat{R}$-module $\widehat{S_1}$. Taking the degree 1 part in the above sequence, we have the exact sequences:

(16.12.1) $\qquad 0 \longrightarrow M \longrightarrow \hat{R}^{(3)} \xrightarrow{(x_1x_3\ x_1x_4\ x_2)} \widehat{S_1} \longrightarrow 0,$

(16.12.2) $\qquad 0 \longrightarrow \hat{R} \xrightarrow{\begin{pmatrix} x_2 \\ -x_4 \\ x_3 \end{pmatrix}} \widehat{S_1} \oplus \widehat{S_{-1}}^{(2)} \longrightarrow M \longrightarrow 0.$

Taking the canonical dual of (16.12.2), we obtain the exact sequence by (16.7):

(16.12.3) $\qquad 0 \longrightarrow M' \longrightarrow \widehat{S_{-2}} \oplus \hat{R}^{(2)} \xrightarrow{(x_2\ -x_4\ x_3)} \widehat{S_{-1}} \longrightarrow 0.$

On the other hand, there is an exact sequence of S-modules:
(16.12.4)

$$0 \longrightarrow S(1) \xrightarrow{\begin{pmatrix} x_4 \\ -x_3 \\ x_2 \end{pmatrix}} S^{(2)} \oplus S(2) \xrightarrow{\begin{pmatrix} x_3 & x_4 & 0 \\ -x_2 & 0 & x_4 \\ 0 & -x_2 & -x_3 \end{pmatrix}} S(-1) \oplus S(1)^{(2)} \xrightarrow{(x_2\ x_3\ x_4)} S$$

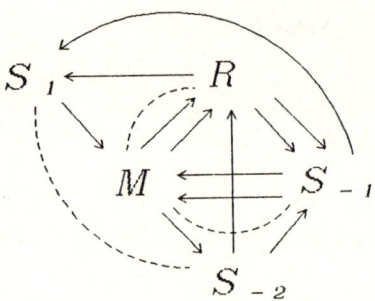

Figure (16.12.6).

This, together with (16.12.3), gives the exact sequence:

$$0 \longrightarrow \widehat{R} \xrightarrow{\begin{pmatrix} x_4 \\ x_3 \\ x_2 \end{pmatrix}} \widehat{S_{-1}}^{(2)} \oplus \widehat{S_1} \longrightarrow M' \longrightarrow 0.$$

Comparing this with (16.12.2) we show that $M \simeq M'$. Thus $\tau(\widehat{S_{-1}}) \simeq (\mathrm{syz}^1_{\widehat{R}}(\widehat{S_1}))' \simeq M' \simeq M$. We have shown:

(16.12.5) $\qquad \tau(\widehat{S_1}) \simeq \widehat{S_{-2}}, \quad \tau(\widehat{S_{-2}}) \simeq \widehat{S_1}, \quad \tau(\widehat{S_{-1}}) \simeq M \quad \text{and} \quad M' \simeq M.$

Applying the functor $\mathrm{Hom}_{\widehat{R}}(\widehat{S_{-1}}, \ )$ to (16.12.3) and noting that $M \simeq M'$, we have an exact sequence

$$\widehat{S_{-1}} \oplus \widehat{S_1}^{(2)} \xrightarrow{(x_2 \ x_4 \ x_3)} \widehat{R} \longrightarrow \mathrm{Ext}^1_{\widehat{R}}(\widehat{S_{-1}}, M),$$

where one can see that $1 \in \widehat{R}$ goes to the socle of $\mathrm{Ext}^1_{\widehat{R}}(\widehat{S_{-1}}, M)$ by the last map. Hence (16.12.3) is an AR sequence. Dually, the sequence (16.12.2) is also an AR sequence. It follows from this that $\tau(M) \simeq \widehat{R}$.

From the Koszul complex over S:

$$0 \longrightarrow S(-3) \xrightarrow{\begin{pmatrix} x_2 \\ -x_1 \end{pmatrix}} S(-2) \oplus S(-1) \xrightarrow{(x_1 \ x_2)} S \longrightarrow S/(x_1, x_2)S \longrightarrow 0,$$

we have an exact sequence:

$$0 \longrightarrow \widehat{S_{-2}} \longrightarrow \widehat{S_{-1}} \oplus \widehat{R} \longrightarrow \widehat{S_1} \longrightarrow 0.$$

In the same manner as above we can show that this sequence is actually an AR sequence. Finally, since (16.12.2) and (16.12.3) are AR sequences, we have irreducible maps $\widehat{S_1} \to M$ and $M \to \widehat{S_{-2}}$, hence the AR sequence ending in $\widehat{S_{-2}}$ must be of the form: $0 \to \widehat{S_1} \to M \to \widehat{S_{-2}} \to 0$.

Thus we have shown that Figure (16.12.6) gives a subgraph of the AR quiver of $\widehat{R}$ that is closed under AR sequences. Hence by (16.9) it is the whole quiver. ∎

(16.13) DEFINITION. Let $a_1 \geq a_2 \geq \ldots \geq a_r > 0$ be given integers and let $\{X_{ij} | 0 \leq j \leq a_i, 1 \leq i \leq r\}$ be a set of variables over k. Then, take the ideal I of the polynomial ring $S_1 = k[X_{ij} | 0 \leq j \leq a_i, 1 \leq i \leq r]$ generated by all the 2 × 2-minors of the matrix:

$$\begin{pmatrix} X_{10} & X_{11} & \cdots & X_{1a_1-1} & | & X_{20} & \cdots & X_{2a_2-1} & | & & | & X_{r0} & X_{r1} & \cdots & X_{ra_r-1} \\ X_{11} & X_{12} & \cdots & X_{1a_1} & | & X_{21} & \cdots & X_{2a_2} & | & \cdots & | & X_{r1} & X_{r2} & \cdots & X_{ra_r} \end{pmatrix}.$$

We define the graded ring R_1 to be S_1/I with $\deg(X_{ij}) = 1$ for all i, j, and call R_1 the **scroll** of type $(a_1, a_2, \ldots, a_r)$. It is known that R_1 is an integral domain of dimension $r + 1$.

We can show that the scroll R_1 is isomorphic to $R(^t(a_1, a_2, \ldots, a_r, -1, -1))$ as a (non-graded) k-algebra. In fact, since

$$R(^t(a_1, a_2, \ldots, a_r, -1, -1)) = k[x_i x_{r+1}^{a_i-j} x_{r+2}^j | 0 \leq j \leq a_i, 1 \leq i \leq r],$$

we can define a mapping $f : R_1 \to R(^t(a_1, a_2, \ldots, a_r, -1, -1))$ by

$$f(X_{ij}) = x_i x_{r+1}^{a_i-j} x_{r+2}^j \quad (0 \leq j \leq a_i, 1 \leq i \leq r),$$

which can be seen to be a well-defined epimorphism of k-algebras. Comparing the Krull dimensions and noting that they are both integral domains, we see that f is an isomorphism.

In particular, the completions of these two graded rings are the same. Hence we can deduce from (15.1) and from (15.14) that:

(16.13.1) The scroll of type $(a_1, a_2, \ldots, a_r)$ is of finite representation type if and only if the integer matrix $(a_1, a_2, \ldots, a_r, -1, -1)$ is of finite type.

Note that we have shown in (16.12) that the scroll of type $(2,1)$ is of finite representation type.

Next we shall give a certain sufficient condition for integer matrices to be of infinite type, (16.15). For this purpose, we settle the notation first.

(16.14) NOTATION. Let A be the same as in (16.6). We can give the structure of a $\mathbb{Z}^2$-graded ring on $S = k[x_1, x_2, \ldots, x_n]$ by

$$\deg(x_i) = (a_i, 1) \quad \text{for} \quad 1 \leq i \leq n.$$

Denote the part of degree $(c, d) \in \mathbb{Z}^2$ in S by $S_{(c,d)}$. Note that, with the notation in (16.6),

$$S_c = \sum_{d \in \mathbb{Z}} S_{(c,d)} \quad (c \in \mathbb{Z}).$$

In particular, $R(A) = \sum_{d \in \mathbb{Z}} S_{(0,d)}$, which gives the natural grading on $R(A)$ that was already given in (16.1.1).

For any $\mathbb{Z}^2$-graded S-module $M = \sum_{(c,d) \in \mathbb{Z}^2} M_{(c,d)}$, we can regard it as a $\mathbb{Z}$-graded S-module by defining the part of degree $c \in \mathbb{Z}$ to be $M_c = \sum_{d \in \mathbb{Z}} M_{(c,d)}$. Note that each M_c is a graded $R(A)$-module. We denote by $M(c, d)$ the shifted $\mathbb{Z}^2$-graded module by (c, d), that is, $M(c, d)_{(c',d')} = M_{(c+c', d+d')}$. Notice that, for any $\mathbb{Z}^2$-graded S-module M and N, the module $\text{Ext}^i_S(M, N)$ is also $\mathbb{Z}^2$-graded, and that

$$\text{Ext}^i_S(M(c, d), N(c', d')) \simeq \text{Ext}^i_S(M, N)(c' - c, d' - d).$$

(16.15) LEMMA. *Given an integer matrix* $A = {}^t(a_1, a_2, \ldots, a_n)$, *suppose that there exist integers* a, b *and* c *such that*

(16.15.1) $\dim_k \text{Ext}^1_S(S_a S, S)_{(-a+b, c)} \geq 2$,

(16.15.2) S_a *and* S_b *are CM modules over* $R(A)$,

(16.15.3) $\text{ht}(S_a S) \geq 2$, $\text{ht}(S_b S) \geq 2$, *and*

(16.15.4) $S_{(a-b, -c)} = 0$.

Then A is of infinite type.

PROOF: Write R instead of $R(A)$. We first claim that

(16.15.5) $\qquad\qquad\qquad \dim_k \text{Ext}^1_R(S_a(-c), S_b)_0 \geq 2.$

For this, let I be a $\mathbb{Z}^2$-graded ideal of S generated by S_a, i.e. $I = S_a S$. Note that $I_a = S_a$. Take a free cover of I to obtain an exact sequence of S-modules:

(16.15.6) $\qquad\qquad 0 \longrightarrow L \longrightarrow S(-a)^{(d)} \longrightarrow I \longrightarrow 0.$

Thus we have the exact sequence:

(16.15.7) $\qquad S(a+b)^{(d)} \longrightarrow \text{Hom}_S(L, S(b)) \longrightarrow \text{Ext}^1_S(I, S(b)) \longrightarrow 0.$

On the other hand, taking the degree a part of (16.15.6) we see that $0 \to L_a \to R^{(d)} \to I_a = S_a \to 0$ is an exact sequence of R-modules. Taking the dual by S_b of this sequence and comparing it with (16.15.7), we have a commutative diagram of graded R-modules with exact rows:

$$\begin{array}{ccccccc}
\mathrm{Hom}_R(R^{(d)}, S_b) & \longrightarrow & \mathrm{Hom}_R(L_a, S_b) & \longrightarrow & \mathrm{Ext}^1_R(S_a, S_b) & \longrightarrow & 0 \\
{\scriptstyle \rho_1}\downarrow & & {\scriptstyle \rho_2}\downarrow & & {\scriptstyle \rho_3}\downarrow & & \\
S_b^{(d)} & \longrightarrow & \mathrm{Hom}_S(L, S(b))_{-a} & \longrightarrow & \mathrm{Ext}^1_S(I, S(b))_{-a} & \longrightarrow & 0,
\end{array}$$

where ρ_1 and ρ_2 are the maps defined in (16.7.1) and ρ_3 is the map induced by ρ_2. Since ρ_1 is an isomorphism, we see that ρ_3 is injective, hence

$$\dim_k \mathrm{Ext}^1_R(S_a, S_b)_c \geq \dim_k \mathrm{Ext}^1_S(I, S)_{(-a+b,c)} \geq 2.$$

This proves (16.15.5).

For any element $\tau \in \mathrm{Ext}^1_R(S_a(-c), S_b)_0$, we write the corresponding extension of R-modules as

$$0 \longrightarrow S_b \longrightarrow M_\tau \longrightarrow S_a(-c) \longrightarrow 0.$$

Hence, by (16.15.2), M_τ is a CM module over R and has rank 2. Secondly we claim the following:

(16.15.8) For $\tau_1, \tau_2 \in \mathrm{Ext}^1_R(S_a(-c), S_b)_0$, if $M_{\tau_1} \simeq M_{\tau_2}$ up to degree shifting as graded R-modules, then τ_1 and τ_2 are lineally dependent over k.

If this is true, then by (16.15.5), we will have an infinite number of nonisomorphic graded CM modules over R of rank 2, hence $\mathfrak{grC}(R)$ is of infinite representation type. Consequently $\mathfrak{C}(R)$ is of infinite representation type by (15.1).

To prove (16.15.8), suppose that there is an isomorphism $f : M_{\tau_1} \to M_{\tau_2}$ of graded R-modules:

$$\begin{array}{ccccccccc}
0 & \longrightarrow & S_b & \stackrel{\iota_1}{\longrightarrow} & M_{\tau_1} & \stackrel{\pi_1}{\longrightarrow} & S_a(-c) & \longrightarrow & 0 \\
& & & & {\scriptstyle f}\downarrow & & & & \\
0 & \longrightarrow & S_b & \stackrel{\iota_2}{\longrightarrow} & M_{\tau_2} & \stackrel{\pi_2}{\longrightarrow} & S_a(-c) & \longrightarrow & 0,
\end{array}$$

where one can see that f must be of degree 0 as an R-homomorphism by comparing the minimal degree in M_{τ_1} and M_{τ_2}. Hence $\pi_2 \cdot f \cdot \iota_1$ is in $\mathrm{Hom}_R(S_b, S_a(-c))_0 \simeq S_{(a-b,-c)}$, and this is zero by (16.15.4). Therefore f induces the mapping $g : S_b \to S_b$ and $h : S_a(-c) \to S_a(-c)$, both of which are R-homomorphisms of degree 0. Since $\mathrm{End}_R(S_a)_0 \simeq \mathrm{End}_S(S)_{(0,0)} \simeq k$ by (16.7), g is a multiplication map by an element of k. Likewise h is also a multiplication by an element of k. Hence we have $\tau_1 = \alpha \cdot \tau_2$ for some $\alpha \in k$. This proves (16.15.8), hence the lemma. ∎

This lemma may be applied to various examples of integer matrices. For example, we can show:

(16.16) PROPOSITION. (Auslander-Reiten [14]) *The integer matrix*

$$A = {}^t(a_1, a_2, \ldots, a_r, -1, -1) \quad (a_1 \geq a_2 \geq \ldots \geq a_r > 0,\ r \geq 2)$$

is of finite type if and only if A is either ${}^t(1,1,-1,-1)$ or ${}^t(2,1,-1,-1)$. In particular, the scroll of type $(a_1, a_2, \ldots, a_r)$ $(r \geq 2)$ is of finite representation type only when its type is either $(2,1)$ or $(1,1)$.

PROOF: We have shown in (16.12) and after (16.2) that the matrices ${}^t(2,1,-1,-1)$ and ${}^t(1,1,-1,-1)$ are of finite type. To show the converse, suppose that the integer matrix A is of finite type. First we notice from (16.8) that

(16.16.1) S_l is a CM $R(A)$-module if $-\sum_{i=1}^r a_i < l \leq 1$.

Next notice that, for any negative integer l, S_l generates the ideal $(x_{r+1}, x_{r+2})^{-l} S$ in S, since any monomials of degree l in S must contain the variables x_{r+1} and x_{r+2} at least $-l$ times. Therefore the ideal $S_l S$ has height two and has the following free resolution as a $\mathbb{Z}^2$-graded S-module:

$$0 \longrightarrow S(-l+1, l-1)^{(-l)} \longrightarrow S(-l, l)^{(-l+1)} \longrightarrow S_l S \longrightarrow 0.$$

Hence the following sequence is exact:

(16.16.2) $S(l, -l)^{(-l+1)} \longrightarrow S(l-1, -l+1)^{(-l)} \longrightarrow \mathrm{Ext}^1_S(S_l S, S) \longrightarrow 0.$

We claim that

(16.16.3) $a_r = 1.$

To show this, suppose that $u_r \geq 2$, and let $s = \#\{i \mid 1 \leq i \leq r, u_i = u_r\}$. We consider two cases and show the contradiction in each case.

The case of $s \geq 2$. In this case, set $a = -a_r - 1$, $b = 1$ and $c = -1$. We take the $(a_r + 2, -1)$-part in (16.16.2) with $l = -a_r - 1$ to get the exact sequence of k-vector spaces:

$$S^{(a_r+2)}_{(1,a_r)} \longrightarrow S^{(a_r+1)}_{(0,a_r+1)} \longrightarrow \mathrm{Ext}^1_S(S_a S, S)_{(a_r+2,-1)} \longrightarrow 0.$$

Since $\dim_k S_{(1,a_r)} = sa_r$ and since $\dim_k S_{(0,a_r+1)} = s(a_r + 1)$, the above sequence implies

$$\dim_k \operatorname{Ext}_S^1(I, S)_{(a_r+2,-1)} \geq s(a_r + 1)^2 - sa_r(a_r + 2) = s \geq 2.$$

Hence the condition (16.15.1) is verified. Secondly, S_{-a_r-1} and S_1 are CM modules over $R(A)$ by (16.16.1). Thus the condition (16.15.2) is satisfied. Thirdly, $\operatorname{ht}(S_{-a_r-1}S) = 2$ as shown in the above, and $\operatorname{ht}(S_1 S) \geq 2$, since S_1 contains $x_1 x_{r+1}^{a_1-1}$ and $x_2 x_{r+2}^{a_2-1}$. This shows that (16.15.3) holds. Finally, $S_{(-a_r-2,1)} = 0$, showing that (16.15.4) is satisfied. Therefore Lemma (16.15) implies that A is of infinite type, a contradiction.

The case of $s = 1$. In this case, take $a = -a_r - 2$, $b = 1$ and $c = -2$. Then the same argument as above shows that the conditions in (16.15) are all satisfied. (Prove this.) This contradicts the assumption that A is of finite type.

Since we get a contradiction in either case, we have shown that $a_r = 1$.

Next we prove that

(16.16.4) $$\sum_{i=1}^{r} a_i \leq 3.$$

Suppose not. Note, then, that S_{-3} and S_1 are CM modules by (16.16.1). Thus, setting $a = -3$, $b = 1$ and $c = -2$, one can verify all the conditions in (16.15), thus A is of infinite type, a contradiction.

We conclude from (16.16.3) and from (16.16.4) that the possible types of A are $t(1, 1, -1, -1)$, $t(2, 1, -1, -1)$ and $t(1, 1, 1, -1, -1)$. We have to show that the last one is excluded. However, this can be done by completely the same argument as above, if we set $a = -2$, $b = 1$ and $c = -1$. We omit the proof of this, leaving it to the reader. ∎

(16.17) *Example.* By a similar argument to the proof of (16.16), the following matrices are shown to be of infinite type by setting a, b and c as indicated below.

Integer Matrix	a	b	c
$t(3, 2, -2, -1)$	-4	4	-3
$t(3, 2, -2, -2)$	-6	2	-2
$t(3, 3, -2, -1)$	-2	5	-2

(16.18) *Example.* Consider an integer matrix $A = {}^t(a, b, -1, -1, -1)$. Then A is of finite type if and only if one of the following conditions holds:

(i) $a = 1$ and $b \leq 0$;

(ii) $b = 1$ and $a \leq 0$;

(iii) $a \leq 0$ and $b \leq 0$.

In fact, if A is of finite type, then, by (16.4), it is necessary that the matrices $^t(a,-1,-1,-1)$, $^t(b,-1,-1,-1)$ and $^t(a,b,-1,-1)$ are of finite type. The conditions for these matrices being of finite type are known from (16.11) and (16.16), which require either A satisfies one of the above conditions, $A = {}^t(1,1,-1,-1,-1)$ or $A = {}^t(2,1,-1,-1,-1)$. If $A = {}^t(1,1,-1,-1,-1)$, then, by (16.3.1), $R(A) \simeq R(^t(1,1,1,-1,-1))$ which is a scroll of type $(1,1,1)$ and is not of finite representation type; see (16.16). Moreover $^t(2,1,-1,-1,-1)$ is also excluded, because

$$\widehat{R}(^t(2,1,-1,-1,-1)) = k[[x_1 x_3^2, x_1 x_3 x_4, x_1 x_4^2, x_1 x_4 x_5, x_1 x_5^2, x_1 x_5 x_3, x_2 x_3, x_2 x_4, x_2 x_5]]$$

is a Gorenstein ring but not a hypersurface and can never be of finite representation type; see (8.15). Conversely if A satisfies one of the above conditions, then $\widehat{R}(A)$ is regular, hence of finite representation type.

We do not know if, aside from $^t(2,-1,-1,-1)$, $^t(2,1,-1,-1)$, $^t(1,1,-1,-1)$ and other trivial ones like (16.18), there are integer matrices of size $n \times 1$ ($n \geq 4$) of finite type or not.

Chapter 17. Homogeneous CM rings of finite representation type

In this chapter we will give a complete classification of homogeneous CM rings of finite representation type according to Eisenbud-Herzog [26]. We shall show that the list is exhausted by the examples we have shown in the previous chapters.

In the rest k will denote an *algebraically closed field of characteristic* 0. Let R be an N-graded CM k-algebra which is generated in degree one and call such R a **homogeneous CM ring** over k. By definition it is described as:

$$R = \sum_{i=0}^{\infty} R_i = R_0[R_1] \quad \text{with} \quad R_0 = k.$$

Let $\mathfrak{m}$ be the irrelevant maximal ideal of R, i.e. $\mathfrak{m} = \sum_{i=1}^{\infty} R_i$ and let $\hat{R}$ be the $\mathfrak{m}$-adic completion of R. Naturally we are interested in the complete local ring $\hat{R}$ of finite representation type. However, we showed in Chapter 15 that $\hat{R}$ is of finite representation type if and only if $\mathfrak{grC}(R)$ is, where $\mathfrak{grC}(R)$ denotes the category of graded CM modules over R and degree-preserving homomorphisms. Thus we mostly think of homogeneous rings themselves other than completions.

We shall be able to give a classification of homogeneous CM rings with $\mathfrak{grC}(R)$ of finite representation type.

First we recall some numerical aspects of homogeneous CM rings.

(17.1) DEFINITION. Let R be a homogeneous CM ring over k of dimension d and let $M = \sum_{i=0}^{\infty} M_i$ be an N-graded module over R with $\text{depth}_R(M) = \dim(M) = t$. The power series in a variable λ:

$$H_M(\lambda) = \sum_{i=0}^{\infty} (\dim_k M_i) \lambda^i$$

is called the **Hilbert series** of M.

It can be seen that

(17.1.1) $$H_M(\lambda) = \frac{P(\lambda)}{(1-\lambda)^t}$$

for some polynomial $P(\lambda)$ with integral coefficients. To show this very quickly, let $\{x_1, x_2, \ldots, x_t\}$ be a regular sequence on M where each x_i is an element of degree one in R and set $_iM = M/(x_1, x_2, \ldots, x_i)M$ ($0 \le i \le t$) and $\overline{M} = {_tM}$. From the exact sequences:

$$0 \longrightarrow (_{i-1}\overline{M})(-1) \xrightarrow{x_i} {_{i-1}\overline{M}} \longrightarrow {_i\overline{M}} \longrightarrow 0 \quad (1 \le i \le t),$$

we see that
$$H_{_i\overline{M}}(\lambda) = (1-\lambda) H_{_{i-1}\overline{M}}(\lambda) \quad (1 \le i \le t).$$

Subsequent use of this gives

$$H_{\overline{M}}(\lambda) = (1-\lambda)^t H_M(\lambda).$$

Since $\overline{M}$ is an Artinian module, $H_{\overline{M}}(\lambda)$ is a polynomial in λ, so that, putting $P(\lambda) = H_{\overline{M}}(\lambda)$, (17.1.1) follows. Note from this that $H_{\overline{M}}(\lambda)$ is independent of the choice of a regular sequence $\{x_1, x_2, \ldots, x_t\}$.

Writing
$$P(\lambda) = \sum_{i=0}^{s} h_i \lambda^i \quad (h_i \in \mathbb{Z}, \; h_s \ne 0),$$

we call the sequence of integers $(h_0, h_1, \ldots, h_s)$ the **h-vector** of M and denote it by $\underline{h}(M)$. Notice that each h_i is nonnegative, because it is the dimension of the degree i part in $\overline{M}$.

A homogeneous CM ring R is said to be **stretched** (in the sense of Sally) if $\underline{h}(R)$ is of the form $(1, n, 1, 1, \ldots, 1)$.

First we remark the following fact:

(17.2) LEMMA. (Eisenbud-Stanley) *Let R be a homogeneous CM ring that is an integral domain with $\underline{h}(R) = (h_0, h_1, \ldots, h_s)$. Then there are inequalities:*

$$h_0 + h_1 + \ldots + h_i \le h_s + h_{s-1} + \ldots + h_{s-i} \quad \text{for} \quad 0 \le i \le [\tfrac{s}{2}].$$

PROOF: Let K_R be the graded canonical module of R and let a be an integer satisfying $(K_R)_a \ne 0$ and $(K_R)_i = 0$ ($i < a$). Shifting the degree of K_R by a, we obtain an $\mathbb{N}$-graded module M with $M_i = (K_R)_{a+i}$ for any i. It is known by Stanley (cf. [62]) that the Hilbert series of M can be written:

$$(17.2.1) \qquad H_M(\lambda) = \frac{h_s + h_{s-1}\lambda + \cdots + h_0\lambda^s}{(1-\lambda)^d}.$$

(Prove this as follows: Reducing modulo a system of parameters, we see that $\overline{M}$ is the canonical module of $\overline{R}$. Then, since $\overline{R}$ is an Artinian local, $\overline{M}$ is the unique indecomposable injective module over $\overline{R}$, hence there is a perfect pairing $\overline{R} \times \overline{M} \to k$. Hence $\dim_k(\overline{R}_i) = \dim_k(\overline{M}_{s-i})$ ($0 \leq i \leq s$), which shows (17.2.1).)

Taking an element $x (\neq 0)$ in M_0, define a degree-preserving homomorphism $\varphi : R \to M$ by $\varphi(r) = rx$ ($r \in R$). Furthermore denote the cokernel of φ by N. Since R is an integral domain and since M is torsion free, we see that φ is a monomorphism, so that the following sequence is exact:

$$(17.2.2) \qquad 0 \longrightarrow R \xrightarrow{\varphi} M \longrightarrow N \longrightarrow 0.$$

Therefore we have

(17.2.3)
$$\begin{aligned}
H_N(\lambda) &= H_M(\lambda) - H_R(\lambda) \\
&= \frac{h_s + h_{s-1}\lambda + \cdots + h_0\lambda^s}{(1-\lambda)^d} - \frac{h_0 + h_1\lambda + \cdots + h_s\lambda^s}{(1-\lambda)^d} \\
&= \frac{\sum_{i=0}^{s}\{(h_s + h_{s-1} + \cdots + h_{s-i}) - (h_0 + h_1 + \cdots + h_i)\}\lambda^i}{(1-\lambda)^{d-1}}.
\end{aligned}$$

(Check the last equality!) On the other hand, we see from (17.2.2) that N is an N-graded module with $\mathrm{depth}(N) = \dim(N) = d-1$, therefore the h-vector of N is nonnegative as remarked before. Hence all the coefficients in the numerator of (17.2.3) are nonnegative. ∎

Applying the lemma to stretched CM rings we have:

(17.3) COROLLARY. *Let R be a stretched homogeneous CM domain. Then the h-vector $\underline{h}(R)$ of R is one of the following:*

$$(1,1,\ldots,1), \quad (1,n) \quad and \quad (1,n,1) \quad with \quad n \geq 2.$$

(17.4) REMARK. Stanley (cf. [62]) showed that the h-vector determines completely the Gorensteinness of a homogeneous CM domain. To be precise, *a homogeneous CM domain R is Gorenstein if and only if the h-vector $(h_0, h_1, \ldots, h_s)$ of R is symmetric, that is, $h_i = h_{s-i}$ for any i ($0 \leq i \leq s$).*

We shall give a proof of this for convenience of the reader. As in the proof of (17.2), let M be the canonical module of R shifted degree, so that it is N-graded and $M_0 \neq 0$.

Then M has h-vector $(h_s, h_{s-1}, \ldots, h_0)$ by (17.2.1). If R is Gorenstein, then $M \simeq R$ therefore $h_i = h_{s-i}$. Conversely suppose $\underline{h}(R)$ is symmetric. Taking an exact sequence as in (17.2.2) one sees from (17.2.3) that the module N has null Hilbert series, therefore $N = 0$ and $R \simeq M$, which means R is Gorenstein.

Combining this remark with (17.3) we proved:

(17.5) COROLLARY. *Let R be a stretched homogeneous CM domain which is not Gorenstein. Then the h-vector of R is $(1,n)$ for some $n \geq 2$.*

(17.6) *Examples.*
(17.6.1) Let I be the ideal of $S = k[U, V, W, X, Y, Z]$ generated by all the 2×2-minors of the symmetric matrix:
$$\begin{pmatrix} U & V & W \\ V & X & Y \\ W & Y & Z \end{pmatrix}.$$
We define the homogeneous CM ring R to be S/I, where all variables have degree one. Then the h-vector is $(1,3)$, hence R is stretched. Note that R is isomorphic to $R(^t(2, -1, -1, -1))$ as a non-graded k-algebra; see (16.10). In particular, R is of finite representation type.

(17.6.2) Let R be the scroll of type $(2,1)$; see (16.13). Then $\underline{h}(R) = (1,2)$ and R is stretched, too. Note that this is also of finite representation type.

As we expect from the above, there is an intimate relation between finiteness of representation type and stretchedness.

(17.7) THEOREM. (Eisenbud-Herzog [26]) *Homogeneous CM rings of finite representation type are stretched.*

PROOF: Suppose that a homogeneous CM ring R is not stretched. We shall construct an infinite number of nonisomorphic indecomposable CM modules over R. To this end, let $\{x_1, x_2, \ldots, x_d\}$ be a system of parameters of R with $\deg(x_i) = 1$ for any i, and let us write $\overline{R} = R/(x_1, x_2 \ldots, x_d)R$ as in (17.1). Since R is not stretched, there is an integer l with $l \geq 2$ and $\dim_k \overline{R}_l \geq 2$. Take $\overline{y} \in \overline{R}_l$. Let $L^\cdot$ be a minimal free resolution of the R-module $\overline{y}\overline{R}$ and let $K^\cdot$ be the Koszul complex of the regular sequence $\{x_1, x_2, \ldots, x_d\}$ that is a free resolution of $\overline{R}$ as an R-module. Since there is a natural embedding $\overline{y}\overline{R} \subset \overline{R}$, we have a morphism $\varphi : L^\cdot \to K^\cdot$ of complexes:

$$\begin{array}{ccccccccccc} \cdots & \to & L^{d+1} & \to & L^d & \to & L^{d-1} & \to & \cdots & \to & L^1 & \to & L^0 & \to & 0 \\ & & & & \varphi^d \downarrow & & \varphi^{d-1} \downarrow & & & & \varphi^1 \downarrow & & \varphi^0 \downarrow & & \\ & & 0 & \to & K^d & \to & K^{d-1} & \to & \cdots & \to & K^1 & \to & K^0 & \to & 0. \end{array}$$

Note that $K^i = R(-i)^{\binom{d}{i}}$. Furthermore, since $\overline{y}$ has degree $l \geq 2$, $L^0 = R(-l)$ and $L^i = \oplus_j R(-a_{ij})$ where $a_{ij} \geq i + l \geq i + 2$. In particular, note that:

(17.7.1) $$(L^{d-1})_m = (L^d)_m = 0 \quad \text{if} \quad m \leq d.$$

Let $C^{\cdot}$ be the mapping cone of $\varphi : L^{\cdot} \to K^{\cdot}$. Since there is an exact sequence of complexes:

$$0 \longrightarrow K^{\cdot} \longrightarrow C^{\cdot} \longrightarrow L^{\cdot}[-1] \longrightarrow 0,$$

taking homologies of the complexes in this sequence, we can easily see that $C^{\cdot}$ is a free resolution of the R-module $\overline{R}/\overline{y}R$. By the definition of mapping cones, $C^{\cdot}$ is described as follows:
(17.7.2)
$$\cdots \to L^d \xrightarrow{\delta^{d+1}} R(-d) \oplus L^{d-1} \xrightarrow{\delta^d} R(-d+1)^{(d)} \oplus L^{d-2} \to \cdots \to R(-1) \oplus L^0 \xrightarrow{\delta^1} R \to 0.$$

Now let $M(\overline{y})$ be the cokernel of δ^{d+1} in this sequence. Since $C^{\cdot}$ is acyclic, $M(\overline{y})$ is a d-th syzygy, hence it is a CM module over R, (1.16). Taking the degree d part in the exact sequence:

$$L^d \to R(-d) \oplus L^{d-1} \to M(\overline{y}) \to 0,$$

we obtain from (17.7.1) $\dim_k M(\overline{y})_d = 1$ and $M(\overline{y})_m = 0$ if $m < d$. Thus there is a unique indecomposable summand $N(\overline{y})$ of $M(\overline{y})$ such that $M(\overline{y})_d = N(\overline{y})_d$. We shall prove:

(17.7.3) For $\overline{y}, \overline{y}' \in \overline{R}_l$, if $N(\overline{y}) \simeq N(\overline{y}')$, then $\overline{y}R = \overline{y}'R$.

If this is true, then, since $\dim_k \overline{R}_l \geq 2$, there are an infinite number of nonisomorphic $N(\overline{y})$'s, hence the proof will be finished.

To prove (17.7.3) let $e(\overline{y})$ be a generator of $N(\overline{y})_d$ as a k-vector space and let $\overline{e}(\overline{y})$ be the natural image of $e(\overline{y})$ in $\overline{N}(\overline{y}) = N(\overline{y})/(x_1, x_2, \ldots, x_d)N(\overline{y})$. Notice that $e(\overline{y})$ is an element in $N(\overline{y})$ of minimal degree and that it is unique up to multiplication by a nonzero element of k. We can show:

(17.7.4) The annihilator $\operatorname{Ann}_R(\overline{e}(\overline{y}))$ is equal to the ideal $(x_1, x_2, \ldots, x_d, y)$, where y is a representative of $\overline{y}$ in R.

Note that (17.7.3) is immediate from this, because, if $N(\overline{y}) \simeq N(\overline{y}')$, then $e(\overline{y})$ is sent to $e(\overline{y}')$ (up to multiplication by a nonzero element in k) by this isomorphism, hence $\operatorname{Ann}_R(\overline{e}(\overline{y})) = \operatorname{Ann}_R(\overline{e}(\overline{y}'))$.

For the proof of (17.7.4), we compute $\operatorname{Tor}_d^R(\overline{R}/\overline{y}\overline{R}, \overline{R})$ using the free resolution (17.7.2) of $\overline{R}/\overline{y}\overline{R}$, and get the following exact seqeunce:

$$0 \longrightarrow \operatorname{Tor}_d^R(\overline{R}/\overline{y}\overline{R}, \overline{R}) \xrightarrow{\psi} \overline{M(\overline{y})} \otimes_R \overline{R} \longrightarrow C^{d-1} \otimes_R \overline{R}.$$

Here, clearly, $\operatorname{Tor}_d^R(\overline{R}/\overline{y}\overline{R}, \overline{R}) \simeq (\overline{R}/\overline{y}\overline{R})(-d)$, and the k-subspace $(\overline{R}/\overline{y}\overline{R})(-d)_d$ in this module is exactly mapped onto $\overline{M(\overline{y})}_d$ by ψ. It, hence, follows that the R-submodule of $\overline{M(\overline{y})}$ generated by $\overline{e}(\overline{y})$ is isomorphic to $\overline{R}/\overline{y}\overline{R}$ up to degree shifting, therefore its annihilator is $(x_1, x_2, \ldots, x_d, y)$. ∎

Combining this theorem with (17.5) we show:

(17.8) COROLLARY. *Suppose that a homogeneous CM ring R is of finite representation type. If R is not Gorenstein and if it is an integral domain, then its h-vector is $(1, n)$ for some $n \geq 2$.*

Homogeneous CM domains with h-vector of the form $(1, n)$ were classified by Bertini. They are:

(*i*)   hypersurfaces $k[X_1, X_2, \ldots, X_n]/(Q)$ for some quadratic polynomial Q,

(*ii*) the ring defined in (17.6.1), and

(*iii*) the scrolls defined in (16.13).

Since we already know which, among the above, are of finite representation type, we derive from (17.8):

(17.9) COROLLARY. *Let R be a homogeneous CM domain of finite representation type. If R is not Gorenstein, then R is either the ring defined in (17.6.1) or the scroll of type $(2, 1)$ or type (m) for some m.*

Now we come to the complete classification of homogeneous CM rings of finite representation type which is due to Eisenbud and Herzog.

Let R be a homogeneous CM ring of finite representation type.

First consider the case of $\dim(R) = 0$. Observe that R is a homomorphic image of the polynomial ring in one variable, for otherwise, R would contain infinite distinct ideals I_n ($n \in \mathbb{N}$) and R/I_n would be all nonisomorphic modules. (Notice that all modules over a ring of dimension 0 are CM.) It then follows that $R \simeq k[X]/(f)$ for some polynomial f, but $f = X^m$ for some $m > 0$, because it must be a homogeneous polynomial.

Next consider the case of $\dim(R) \geq 1$. If R is a Gorenstein ring, then by (8.15) it is a hypersurface: $R = k[X_1, X_2, \ldots, X_n]/(f)$. Since $n \geq 2$, f must be the one listed in (8.8). Among them, homogeneous ones are $X_1^2 + X_2^2 + \ldots + X_n^2$ and $X_1^2 X_2 + X_2^3$. Suppose R is not Gorenstein. If $\dim(R) = 1$, then, by virtue of the graded version of (9.2), R birationally dominates a simple hypersurface singularity that is homogeneous. Such a non-Gorenstein ring is easily seen to be isomorphic to the ring $k[X,Y,Z]/(XY,YZ,ZW)$, cf. (9.16). If $\dim(R) \geq 2$, then R is an integral domain, because it is an isolated singularity, (4.22). In this case, we have already done by the above corollary.

Summing up the above, we have shown:

(17.10) THEOREM. (Eisenbud-Herzog [26]) *A homogeneous CM ring of finite representation type is isomorphic to one of the following rings:*

(i) $k[X_1, X_2, \ldots, X_n]$,

(ii) $k[X]/(X^m)$ *for some* $m \geq 1$,

(iii) $k[X_1, X_2, \ldots, X_n]/(X_1^2 + X_2^2 + \ldots + X_n^2)$,

(iv) $k[X_1, X_2]/(X_1^2 X_2 + X_2^3)$,

(v) $k[X_1, X_2, X_3]/(X_1 X_2, X_2 X_3, X_3 X_1)$,

(vi) *the scroll of type* (m) *for some* m,

(vii) *the scroll of type* $(2,1)$, *and*

(viii) *the ring defined in* (17.6.1).

Addenda

(*A*.1) We were not able to discuss geometric aspect of CM modules.
To give a brief explanation on this, let R be a complete two-dimensional normal local domain over an algebraically closed field of characteristic 0. One can construct a minimal desingularization $\pi : Y \to X = \mathrm{Spec}(R)$. Let E be the closed fiber of π and decompose it as $E = \bigcup_i E_i$ where each E_i is an irreducible component of E. Suppose that R is a simple hypersurface singularity. Then it is known that all the E_i are rational curves. For a CM module M over R, we define

$$\widetilde{M} = \pi^* M/\mathfrak{m}\text{-torsion},$$

which is known to be a vector bundle on Y. If M is indecomposable and nonfree, then there is a unique i such that

$$(c_1(\widetilde{M}) \cdot E_i) = 1,$$

where $c_1(\widetilde{M})$ is the first Chern class of $\widetilde{M}$. Artin and Verdier [3] showed that this correspondence $M \mapsto E_i$ gives rise to a bijection between the set of classes of indecomposable, nonfree CM modules over R and the set $\{E_i\}$. This explains how the Dynkin diagrams appear as AR quivers of R. See also Esnault [27], Esnault-Knörrer [28] and Kahn [42] for further discussion in this direction.

(*A*.2) There is a noncommutative analogy of CM modules.
Let Λ be an order in an Azumaya algebra over a field and let R be its center. A left Λ-module is said to be CM if it is a CM module as an R-module. The problem is to classify orders which have only a finite number of isomorphism classes of indecomposable CM modules. In [1], Artin succeeded in doing this in the case that R is a complete regular local ring of dimension 2 of equicharacteristic 0 and that Λ is a maximal R-order. For two-dimensional maximal orders with arbitrary centers of characteristic 0, it was also done by Artin [2]. See also Reiten-Van den Bergh [54] for non-maximal orders.

(A.3) A module M over a local ring R is said to be a **Buchsbaum module** if, for any system of parameters $\{x_1, x_2, \ldots, x_d\}$ of M,

$$(x_1, x_2, \ldots, x_i)M : x_{i+1} = (x_1, x_2, \ldots, x_i)M : \mathfrak{m} \quad (0 \leq i \leq d-1),$$

where $\mathfrak{m}$ is the maximal ideal of R. Furthermore, M is a **maximal Buchsbaum module** if it is a Buchsbaum module with $\dim(M) = \dim(R)$. Clearly, CM modules are maximal Buchsbaum. Goto proved that a regular local ring has only a finite number of isomorphism classes of indecomposable maximal Buchsbaum modules. Again a question arises: When does R have only a finite number of classes of indecomposable maximal Buchsbaum modules ?

In [30], Goto and Nishida showed that, if R is a complete CM local ring of dimension ≥ 2 and if $R/\mathfrak{m}$ is an algebraically closed field of characteristic unequal to 2, then such finiteness leads to the regularity of R.

References

1. M.Artin, *Maximal orders of global dimension and Krull dimension two*, Invent. Math. **84** (1986), 195–222; *Correction*, Invent. Math. **90** (1987), p. 217.
2. M.Artin, *Two dimensional orders of finite representation type*, Manuscripta Math. **58** (1987), 445–471.
3. M.Artin and J.-L.Verdier, *Reflexive modules over rational double points*, Math. Ann. **270** (1985), 79–82.
4. M.Auslander, *Representation theory of artin algebras I*, Comm. Alg. **1** (1974), 177–268; *II*, Comm. Alg. **2** (1974), 269–310.
5. M.Auslander, *Functors and morphisms determined by objects*, Proc. Conf. Representation Theory, Philadelphia 1976, Marcel Dekker (1978), 1–224.
6. M.Auslander, *Rational singularities and almost split sequences*, Trans. Amer. Math. Soc. **293** (1986), 511–531.
7. M.Auslander, *Isolated singularities and existence of almost split sequences*, Proc. ICRA IV, Springer Lecture Notes in Math. **1178** (1986), 194–241.
8. M.Auslander and M. Bridger, "Stable module theory," Mem. Amer. Math. Soc. vol. 94, 1969.
9. M.Auslander and I.Reiten, *Representation theory of artin algebras III*, Comm. Alg. **3** (1975), 239–294; *IV*, Comm. Alg. **5** (1977), 443–518.
10. M.Auslander and I.Reiten, *Grothendieck groups of algebras and orders*, J. Pure Appl. Algebra **39** (1986), 1–51.
11. M.Auslander and I.Reiten, *McKay quivers and extended Dynkin diagrams*, Trans. Amer. Math. Soc. **293** (1986), 293–301.
12. M.Auslander and I.Reiten, *Almost split sequences for Z-graded rings*, Singularities, representation of algebras and vector bundles (Lambrecht,1985), Springer Lecture Notes in Math. **1273** (1987), 232–243.
13. M.Auslander and I.Reiten, *Almost split sequences in dimension two*, Adv. in Math. **66** (1987), 88–118.
14. M.Auslander and I.Reiten, *The Cohen-Macaulay type of Cohen-Macaulay rings*, Adv. in Math. **73** (1989), 1–23.
15. M.Auslander and I.Reiten, *Cohen-Macaulay modules for graded Cohen-Macaulay rings and their completions*, Commutative Algebra, Proceedings of a Micro Program (Berkeley, 1987), Springer Lecture Notes in Math. (1989), 21–31.
16. N.Bourbaki, "Algèbre Commutative, Chapter 7," Hermann, Paris, 1968.

17. N.Bourbaki, "Groupe et Algèbre de Lie, Chapter 5," Hermann, Paris, 1968.
18. R.-O.Buchweitz, D.Eisenbud and J.Herzog, *Cohen-Macaulay modules on quadrics*, Singularities, representation of algebras and vector bundles (Lambrecht,1985), Springer Lecture Notes in Math. **1273** (1987), 58–116.
19. R.-O.Buchweitz, G.M.Greuel and F.-O.Schreyer, *Cohen-Macaulay modules on hypersurface singularities II*, Invent. Math. **88** (1987), 165–182.
20. M.C.R.Butler, *Grothendieck groups and almost split sequences*, Springer Lecture Note in Math. **882** (1981), 357–368.
21. E.Dieterich, *Reduction of isolated singularities*, Comment. Math. Helv. **62** (1987), 654–676.
22. E.Dieterich, *Auslander-Reiten quiver of an isolated singularity*, Singularities, representation of algebras and vector bundles (Lambrecht,1985), Springer Lecture Notes in Math. **1273** (1987), 244–264.
23. E.Dieterich and A.Wiedemann, *Auslander-Reiten quiver of a simple curve singularity*, Trans. Amer. Math. Soc. **294** (1986), 455–475.
24. J.A.Drozd and A.V.Roiter, *Commutative rings with a finite number of indecomposable integral representations*, Math. USSR Izv. **6** (1967), 757–772.
25. D.Eisenbud, *Homological algebra on a complete intersection, with an application to group representations*, Trans. Amer. Math. Soc. **260** (1980), 35–64.
26. D.Eisenbud and J.Herzog, *The classification of homogeneous Cohen-Macaulay rings of finite representation type*, Math. Ann. **280** (1988), 347–352.
27. H.Esnault, *Reflexive modules on quotient singularities*, J. Reine Angew. Math. **362** (1985), 63–71.
28. H.Esnault and H.Knörrer, *Reflexive modules over rational double points*, Math. Ann. **272** (1985), 545–548.
29. E.G.Evans and P.Griffith, "Syzygies," London Math. Soc., Lecture Note Series vol. 106, Cambridge U. P., 1985.
30. S.Goto and K.Nishida, *Rings with only a finite number of isomorphism classes of indecomposable maximal Buchsbaum modules*, J. Math. Soc. Japan **40** (1988), 501–518.
31. E.L.Green and I.Reiner, *Integral representations and diagrams*, Michigan Math. J. **25** (1978), 53–84.
32. G.M.Greuel and H.Knörrer, *Einfache Kurvensingularitäten und torsionsfreie Moduln*, Math. Ann. **270** (1985), 417–425.
33. A.Grothendieck, "Local Cohomology," Lecture Notes in Math. vol. 41, Springer Verlag, 1967.
34. M.Harada and Y.Sai, *On categories of indecomposable modules I*, Osaka J. Math. **8** (1971), 309–321.

35. J.Herzog, *Ringe mit nur endlich vielen Isomorphieklassen von maximalen unzerlegbaren Cohen-Macaulay-Moduln*, Math. Ann. **233** (1978), 21–34.
36. J.Herzog, *Linear Cohen-Macaulay modules on integral quadrics*, Séminaire d'algèbre P.Dubreil et M.-P.Malliavin, Springer Lecture Notes in Math. **1296** (1987), 214–227.
37. J.Herzog and E.Kunz, "Der kanonische Modul eines Cohen-Macaulay-Rings," Lecture Notes in Math. vol. 238, Springer Verlag, 1971.
38. J.Herzog and H.Sanders, *Indecomposable syzygy-modules of high rank over hypersurface rings*, J. Pure Appl. Algebra **51** (1988), 161–168.
39. G.Hochschild, *Cohomology groups of an associative algebra*, Ann. Math. **46** (1945), 58–67.
40. M.Hochster, *Rings of invariants of tori, Cohen-Macaulay rings generated by monomials*, Ann. Math. **96** (1972), 318–337.
41. H.Jacobinski, *Sur les ordres commutatifs avec un nombre fini de réseaux indecomposables*, Acta Math. **118** (1976), 1–31.
42. C.Kahn, *Reflexive Moduln auf einfach-elliptischen Flächensingularitäten*, Brandeis Univ. (1987).
43. K.Kiyek and G.Steinke, *Einfache Kurvensingularitäten in beliebiger Charakteristik*, Arch. Math. **45** (1985), 565–573.
44. H.Knörrer, *Cohen-Macaulay modules on hypersurface singularities I*, Invent. Math. **88** (1987), 153–164.
45. H.Knörrer, *Cohen-Macaulay modules on hypersurface singularities*, Representation of algebras, London Math. Soc. **116** (1985), 147–164.
46. T.Y.Lam, "The Algebraic Theory of Quadratic Forms," Mathematics Lecture Note Series, W.A.Benjamin, 1973.
47. H.Matsumura, "Commutative Algebra," Benjamin, 2nd edition, Mass., 1980.
48. H.Matsumura, "Commutative Ring Theory," Cambridge Studies in Advanced Math. vol. 8, Cambridge U.P., 1989.
49. D.Mumford, *The topology of normal singularities of an algebraic surface and a criterion for simplicity*, Publ. Math. IHES **9** (1961), 5–22.
50. M.Nagata, "Local Rings," Krieger, Huntington, 1975.
51. R.S.Pierce, "Associative Algebras," Graduate Texts in Math. vol. 88, Springer Verlag, 1982.
52. D.Popescu, *Indecomposable Cohen-Macaulay modules and their multiplicities*, Trans. Amer. Math. Soc. (to appear).
53. I.Reiten, *Finite-dimensional algebras and singularities*, Singularities, representation of algebras and vector bundles (Lambrecht,1985), Springer Lecture Notes in Math. **1273** (1987), 35–57.

54. I.Reiten and M.Van den Bergh, *Two-dimensional tame orders and maximal orders of finite representation type*, The University of Trondheim (1987).
55. C.M.Ringel, *Report on the Brauer-Thrall conjectures*, Proc. ICRA I, Ottawa, Springer Lecture Notes in Math. **831** (1980), 104–136.
56. A.V.Roiter, *Unbounded dimensionality of indecomposable representations of an algebra with an infinite number of indecomposable representations*, Math. USSR Izv. **2** (1968), 1223–1230.
57. G.Scheja and U.Storch, "Lokale Verzweigungstheorie," Schriftenreihe des Math. Inst. der Univ. Freiburg, 1974.
58. F.-O.Schreyer, *Finite and countable CM representation type*, Singularities, representation of algebras and vector bundles (Lambrecht,1985), Springer Lecture Notes in Math. **1273** (1987), 9–34.
59. J.-P.Serre, "Local Fields," Graduate Texts in Math. vol. 67, Springer Verlag, 1979.
60. Ø.Solberg, *Hypersurface singularities of finite Cohen-Macaulay type*, Proc. London Math. Soc. **58** (1989), 258–280.
61. Ø.Solberg, *A graded ring of finite Cōhen-Macaulay type*, Comm. Alg. **16** (1988), 2121–2124.
62. R.P.Stanley, "Combinatorics and Commutative Algebra," Progress in Math. vol. 41, Birkhäuser, 1983.
63. J.Tate, *Homology of Noetherian rings and local rings*, Illinois J. Math. **1** (1957), 14–25.
64. K.Watanabe, *Certain invariant subrings are Gorenstein, I*, Osaka J. Math. **11** (1974), 1–8; *II*, 379–388.
65. J.Wunram, *Reflexive modules on cyclic quotient surface singularities*, Singularities, representation of algebras and vector bundles (Lambrecht, 1985), Springer Lecture Notes in Math. **1273** (1987), 221–231.
66. Y.Yoshino, *Brauer-Thrall type theorem for maximal Cohen-Macaulay modules*, J. Math. Soc. Japan **39** (1987), 719–739.
67. Y.Yoshino, "CM modules over Hensel CM rings (in Japanese)," Lecture Notes Series of Math., Tokyo Metropolitan Univ., 1987.
68. Y.Yoshino and T.Kawamoto, *The fundamental modules of a normal local domain of dimension 2*, Trans. Amer. Math. Soc. **309** (1988), 425–431.

Index

admitting AR sequences, 15
algebraic torus, 143
analytic algebra, 6
AR quiver, 36
AR sequence, 10, 14
AR translation, 10
Auslander category, 26
Auslander transpose, 18
Betti number, 68
birationally dominate, 69
bounded multiplicity type, 44
Brauer-Thrall theorem, 45
Buchsbaum module, 169
canonical module, 3
chain of irreducible morphisms, 51
Clifford algebra, 130
closed under AR sequences, 149
CM approximation, 33
Cohen-Macaulay module (CM module), 1
conductor, 72
convergent power series, 6
Dedekind different, 46
efficient system of parameters, 48
equivalence of matrix factorization, 55
finitely generated functor, 26
finitely presented functor, 26
finite representation type, 39, 135
fundamental module, 100
fundamental sequence, 100
gradable module (homomorphism), 137
graded isolated singularity, 138
Grothendieck group, 118
Harada-Sai lemma, 51
Henselian ring, 5

Hilbert series, 161
homogeneous CM ring, 161
h-vector, 162
hypersurface, 54
indecomposable, 5
integer matrix of (in)finite type, 144
irreducible morphism, 11
isolated singularity, 16
isomorphism up to degree shifting, 135
Klein group, 94
local cohomology, 1
local duality, 4
locally free on punctured spectrum, 17
matrix factorization, 55
McKay graph, 87
minimal reduction, 62
morphism of matrix factorizations, 55
multiplicity, 2
Noetherian different, 46
Noetherian normalization, 3
periodic resolution, 56
periodicity except for a finite part, 68
pseudo-reflection, 88
quadratic space, 124
reduced matrix factorization, 55
reduced syzygy, 5
representation finite, 39
scroll, 155
separable system of parameters, 6, 46
simple object, 28
simple (hypersurface) singularity, 60
split morphism, 6
stable AR quiver, 117
stretched ring, 162
syzygy, 5
toric singularity, 143
Yoneda's lemma, 26

Index of Symbols

$\dim_R M$, the Krull dimension of an R-module M.
$\text{depth}_R M$, the depth of an R-module M.
$\text{length}_R(M)$, the length of an Artinian R-module M.
$H^i_\mathfrak{m}(M)$, i-th local cohomology of M with support in $\{\mathfrak{m}\}$; (1.2).
$e(M)$, the multiplicity of M; (1.6).
K_R, the canonical module of R; (1.10).
$\text{syz}^n(M)$, the reduced n-th syzygy of M; (1.15).
$k\{x_1, x_2, \ldots, x_n\}$, convergent power series ring over a valued field k; (1.19).
$\mathfrak{M}(R)$, the category of finitely generated R-modules.
$\mathfrak{C}(R)$, the category of CM modules over R.
$S(M)$ (resp. $S'(N)$); (2.1) (resp. (2.1)′).
$\tau(M)$, the AR translation of M; (2.8).
$\tau^{-1}(M)$; (2.8)′.
$\text{tr}(M)$, the Auslander transpose of M; (3.5).
$\underline{\text{Hom}}_R(M, N)$; (3.7).
$\text{Mod}(\mathfrak{C})$, the category of contravariant additive functors from $\mathfrak{C}$ to (Ab); (4.1).
$\text{mod}(\mathfrak{C})$, the Auslander category of $\mathfrak{C}$; (4.6).
(M, N), $= \text{Hom}_R(M, N)$ in Chapters 4 and 5.
S_M, the simple functor associated to M; (4.11).
$\underline{\text{Mod}}(\mathfrak{C})$; (4.14).
$\underline{\text{mod}}(\mathfrak{C})$; (4.14).
$(M, N)_n$; (5.1).
$\text{Irr}(M, N)$, the space of irreducible morphisms from M to N; (5.1).
$\text{irr}(M, N)$, the dimension of $\text{Irr}(M, N)$; (5.1).
$n(R)$, the number of isomorphism classes of indecomposable CM modules over R; (5.10).
$\mathcal{N}^R_T$, the Noetherian different of R over T; (6.6).
$\mathcal{D}^R_T$, the Dedekind different of R over T; (6.6).
$H^i_T(R, M)$, i-th Hochschild cohomology of an R-module M; (6.8).
$\mathcal{N}^R$; (6.11).
$MF_S(f)$, the category of matrix factorizations of f; (7.1).
$\text{Coker}(\varphi, \psi)$; (7.2), (7.4).
$\underline{MF}_S(f)$; (7.3).
$\underline{RMF}_S(f)$; (7.3).
$\underline{\mathfrak{C}}$; (7.3).

Index of Symbols

$c(f)$, the set of ideals I with $f \in I^2$; (8.1).
$I(M)$; (8.11).
$\beta_n(M)$, n-th Betti number of M; (8.18).
$S * G$, the skew group ring of G over S; (10.1.1).
$\wp(S * G)$, the category of projective modules over $S * G$; (10.1).
$\mathrm{Mc}(V, G)$, the McKay graph of G on V; (10.3).
$mult_i(W)$; (10.3).
$\nu_i(P)$; (10.3).
$R^\natural$; (12.1.1).
$R^{\natural\natural}$; (12.8).
$\underline{\Gamma}(R)$, the stable AR quiver of R; (12.12).
$\mathrm{K}_0(\mathfrak{A})$, the Grothendieck group of $\mathfrak{A}$; (13.1).
$\mathrm{AR}(\mathfrak{C})$; (13.6).
R_Q, the local ring of hypersurface defined by a quadratic form Q; (14.1).
G_Q, the associated graded ring of R_Q; (14.4).
$C(Q)$, the Clifford algebra of Q; (14.6).
$\mathfrak{gr}\mathfrak{M}(C(Q))$, the category of graded $C(Q)$-modules; (14.7).
$\mathfrak{gr}\mathfrak{C}(R)$, the category of graded CM modules over R; (15.1).
τ_{gr}, graded AR translation; (15.4.1).
Γ_{gr}, the subgraph of Γ consisting of gradable modules; (15.9).
$R(A)$, the semigroup ring defined by an integer matrix A; (16.1).
$\widehat{R}(A)$, the toric singularity associated with an integer matrix A; (16.1).
$H_M(\lambda)$, the Hilbert series of M; (17.1).
$\underline{h}(M)$, the h-vector of M; (17.1).